Power Systems

Electrical power has been the technological foundation of industrial societies for many years. Although the systems designed to provide and apply electrical energy have reached a high degree of maturity, unforeseen problems are constantly encountered, necessitating the design of more efficient and reliable systems based on novel technologies. The book series Power Systems is aimed at providing detailed, accurate and sound technical information about these new developments in electrical power engineering. It includes topics on power generation, storage and transmission as well as electrical machines. The monographs and advanced textbooks in this series address researchers, lecturers, industrial engineers and senior students in electrical engineering.

Power Systems is indexed in Scopus

Ikromjon U. Rakhmonov · Vasiliy Ya. Ushakov ·
Foziljon A. Khoshimov · Numon N. Niyozov ·
Nurbek N. Kurbonov · Alexey V. Mytnikov

Electric Consumption by Industrial Enterprises

Modeling, Rationing and Forecasting

Springer

Ikromjon U. Rakhmonov
Department of Power Supply
Tashkent State Technical University
Tashkent, Uzbekistan

Vasiliy Ya. Ushakov
School of Energy and Power Engineering
Tomsk Polytechnic University
Tomsk, Russia

Foziljon A. Khoshimov
Department of Energy Efficiency
Academy of Sciences of Uzbekistan
Tashkent, Uzbekistan

Numon N. Niyozov
Department of Power Supply
Tashkent State Technical University
Tashkent, Uzbekistan

Nurbek N. Kurbonov
Department of Power Supply
Tashkent State Technical University
Tashkent, Uzbekistan

Alexey V. Mytnikov
School of Energy and Power Engineering
Tomsk Polytechnic University
Tomsk, Russia

ISSN 1612-1287 ISSN 1860-4676 (electronic)
Power Systems
ISBN 978-3-031-62675-3 ISBN 978-3-031-62676-0 (eBook)
https://doi.org/10.1007/978-3-031-62676-0

This Springer imprint is published by the registered company Springer Nature Switzerland AG
The registered company address is: Gewerbestrasse 11, 6330 Cham, Switzerland

Preface

As the threats of depletion of mineral energy resources and the infliction of irreversible environmental damage escalate, coupled with the high cost of electricity produced from renewable energy resources, developed countries are increasingly focusing on developing methods for managing power consumption. This is aimed at enhancing the efficiency of electric power usage by all consumers, especially industrial enterprises. In these countries, the share of industrial enterprises in total power consumption is very significant and continues to grow at the following rates: 54% in 2020, 59% in 2021, and 61% in 2022.

The core of these management methods is the assessment, modeling, standardization, and forecasting of power consumption, considering the complexity of technological processes due to their multi-stage nature. Various research teams are developing models, algorithms, and software products primarily aimed at monitoring the power consumption of enterprises in energy-intensive sectors—such as ferrous and non-ferrous metallurgy, chemical, and textile industries, among others. Priority is given to research and developments that take into account the inter-industrial connections of technological processes in the analysis and calculation of power consumption by industrial enterprises. An ongoing task is the development of methods and algorithms for determining scientifically justified norms, highly accurate forecasts, and software products that allow monitoring the electric power consumption of industrial enterprises and choosing the correct paths to realize the still very large potential for energy saving and efficiency improvement.

Effective measures for the rational use of electricity at enterprises will only be possible when they are subordinated to a system whose main elements are: accounting and control of power consumption, compilation of power balances for individual machines and units, workshops, and the enterprise as a whole, rationing of power consumption, and the development and implementation of specific measures for rational use of electricity.

A significant element of the system is the presence of a base of mobile devices that allows systematic measurements of electricity, heat, compressed air, and other energy carriers consumption using portable devices in those points of the power supply scheme where stationary energy accounting devices are absent for various reasons. Such a base is also necessary for conducting selective tests of equipment (idle power measurements, load diagram records, etc.).

For concentrating the efforts of scientists and specialists in searching for reserves to economize fuel and energy resources, as well as ways to solve specific tasks in implementing energy-saving policies, new methods and technical solutions are being developed to optimize the consumption of fuel, electrical and thermal energy in the main processes of industrial production, using technical means of accounting, control, and management. An important element in solving these tasks is the development of methods for experimental research on existing industrial equipment and the creation of mathematical models of control parameters for energy consumption modes, ensuring minimum energy expenditure.

This monograph was prepared within the framework of the fundamental project FOT-2021–236 on the topic "Research on the patterns of complex management of power consumption by industrial enterprises with continuous production," carried out by scientists of Tashkent State Technical University and Tomsk Polytechnic University.

To ensure better accessibility for readers to the accumulated materials on the problem discussed in the monograph, it contains extensive bibliographies for each chapter. Special attention is paid to those containing materials on the problems of modeling, standardizing, and forecasting the power consumption of industrial enterprises. The authors' own works are also focused on these three problems. As a result, it was possible to formulate scientifically substantiated recommendations for their solution, test and confirm their adequacy in operating industrial enterprises belonging to various sectors of the economy.

The team of authors expresses gratitude to Academician Allaev Kakhramon Rakhimovich of the Academy of Sciences of the Republic of Uzbekistan, who made a number of extremely valuable suggestions for improving the monograph.

Comments and suggestions, which readers can send to the following email addresses, will be gratefully received:

vyush@tpu.ru.; mytnikov66@mail.ru

Tashkent, Uzbekistan	Ikromjon U. Rakhmonov
Tomsk, Russia	Vasiliy Ya. Ushakov
Tashkent, Uzbekistan	Foziljon A. Khoshimov
Tashkent, Uzbekistan	Numon N. Niyozov
Tashkent, Uzbekistan	Nurbek N. Kurbonov
Tomsk, Russia	Alexey V. Mytnikov

About This Book

The monograph contains the results of research and accumulated experience in implementing the concept of energy saving and increasing energy efficiency (ES and IEE) at industrial enterprises that consume a significant share of the electricity produced in industrialized countries. It is carefully structured into eleven chapters that progressively cover the main aspects of this vital concept.

Chapter 1 introduces the reader to the basic concepts of ES and IEE. It emphasizes the importance of optimizing energy saving measures based on instrumental surveys and energy audits of electricity consumers.

Chapter 2 analyzes existing direct and indirect methods of energy management, based on modeling, normalization, and forecasting.

Chapter 3 discusses the structure, types, and areas of application of energy balances. The main stages of compiling actual and ideal balances are described; their great importance in energy management is emphasized.

Chapter 4 gives an enlarged description of industrial enterprises belonging to the main branches of industry. They are integrated into ten groups depending on their production cycles and the types of equipment used.

In Chap. 5, the patterns and problems of energy consumption management in conditions of changing demand are analyzed, and effective methods are proposed for smoothing daily energy consumption schedules in various branches of industry.

Chapter 6 is devoted to solving the problem of increasing the accuracy of calculations of the energy intensity of industrial products, based, among other things, on decomposition and estimation of electricity consumption for own needs.

Chapter 7 considers the factors influencing electricity consumption and presents methods for determining normalized values based on multifactor analysis and cost minimization.

Chapter 8 discusses the composition and classification of electrical energy consumption norms. It substantiates the relevance of the application of consolidated and differentiated standards and proposes methods for determining these norms.

Chapter 9 is devoted to the development of models for forecasting electricity consumption and methods for their estimation. Particular attention is paid to the use of artificial neural networks to improve the accuracy of forecasts.

Chapter 10 is devoted to the analysis of possible ways to save electricity, describing the methodology for estimation the consequences of its impact on the technological and organizational operating conditions of industrial enterprises.

In Chap. 11, using the example of enterprises belonging to various industries, the reduction of specific power consumption by optimizing the parameters of raw materials and semi-finished products is considered.

The monograph is intended for engineering personnel and researchers working in the field of industrial energy management, as well as for undergraduate and graduate students specializing in this area.

Each chapter begins with a brief summary and ends with a list of references.

Contents

Chapter 1
Estimation of the Power Saving Potential and Economic Efficiency of Its Implementation

This chapter names and explains the basic concepts in the field of energy saving and improvement of the energy efficiency (ES and IEE), which are the main subject of consideration in this monograph with an emphasis on industrial enterprises. It is shown that not in all cases the amount of savings obtained as a result of the implementation of measures on ES and IEE exceeds the funds invested in them. Optimization of this activity is necessary. An important tool for this is an energy survey, which includes an instrumental examination of a facility, for example an industrial enterprise, and an energy audit (The study of documents that reflect the consumption of electricity and its payment). It is shown that in order to assess the economic efficiency of measures for ES and IEE, it is necessary to use at least 4 target indicators applicable in the main areas of the economic and social life of society.

1.1 Basic Concepts from the Sphere of Power Saving and Improving the Energy Efficiency

The basic concepts that refer to the problem of reducing specific energy consumption for the production of goods and services can be defined as follows [1–3]:

Energy saving is the implementation of the organizational, legal, technical, technological, economic, and other measures aimed at reducing the relative content of employed energy resources with preservation of the corresponding positive effect from their utilization;

Energy efficiency is the characteristic describing the ratio of the useful effect on the utilization of power resources to the cost of energy resources spend to obtain this effect.

I. U. Rakhmonov et al., *Electric Consumption by Industrial Enterprises*, Power Systems, https://doi.org/10.1007/978-3-031-62676-0_1

In scientific literature *energy saving* is often treated as actions/measures aimed at elimination of wastefulness caused mainly by organizational disorders and energy profligate population behavior and at restoration of elementary order;

Improvement of energy efficiency is taken to mean reduction in the energy intensity of technological processes and industrial and household equipment.

The main indicators of energy efficiency are:

- The efficiency of a power plant,
- The efficiency of energy consumption,
- The efficiency of consumption of separate energy carriers and their components,
- specific (actual) consumption of an energy carrier,
- specific consumption of energy resources per unit of product including individual energy carriers (electric power, fuel, and water).

A need for investments and capital investments in energy saving and improvement of energy efficiency (ES and IEE) requires the feasibility studies of these measures. Very often, it is required to optimize monetary, labor, and other expenses (expenses below) on the introduction of energy saving technologies and energy efficient equipment. Wherein, it is necessary to consider nonlinear dependences of energy saving on investments. Implementation of ES and IEE measures leads on the one hand, to a decrease in the specific energy consumption that causes the product cost to decrease and on the other hand, additional money is required for modernization of energy-consuming industrial equipment that causes the product cost to increase. Qualitatively, Fig. 1.1 explains this.

For example, it is well known that not any decrease in losses of electric energy transportation in an electricity grid increases the economy of its operation as a whole. The losses can also be decreased by economically inexpedient methods. At the same time, not each increase of the cost effectiveness of electricity grid operation is accompanied by decreased energy losses in this grid.

Thus, the problem of substantiation of the ES and IEE measures is reduced to optimization of costs of their implementation. Since measures for improving energy efficiency require much time, this problem is complicated because of the instability

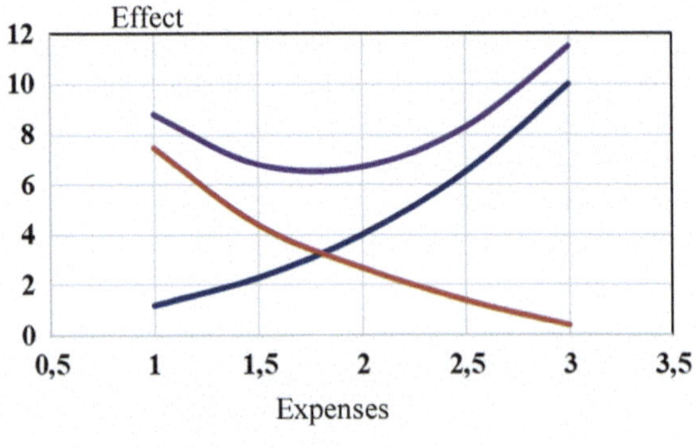

Fig. 1.1 To the explanation of the need for optimization of expenses for ES and IEE measures. Here curve (black) (is for the product cost (In conventional units) with allowance for additional expenses, curve (brown) is for the cost of the consumed energy, and curve (blue) shows the total effect of energy saving measures

of costs. Inflation requires that optimization was performed against the background of predicted development of economy and costs of fuel and energy resources (FER). In addition, a hidden effect from fossil hydrocarbon fuel saving (the environmental component of the effect) is difficult to quantify.

To estimate quantitatively maximum possible results of energy saving activity, the characteristic of **energy saving potential (ESP)** was considered**.** In special literature at least three definitions of this characteristic are given:

(1) The amount of energy resources by which their content in the production of the same volume and nomenclature of products, goods, and services of invariable quality within the preset time can be decreased;
(2) Maximum losses of fuel and thermal, mechanical, and electric energy on the level of a plant, cycle, shop, and factory that can be returned fully or partly into the energy technology cycle using the corresponding energy saving measures;
(3) The difference between the actual consumption of energy resources and generated energy and the progressive norms of energy resource consumption or the expected result (in %) of expense reduction due to energy saving measures undertaken by the consumer. The ESP is estimated by results of energy audit and instrumental energy inspection of energy and energy resources consumers.

Depending on the energy consumption with which the actual energy consumption is compared, the energy saving potential can be:

- theoretical (ideal),
- practical (reference),
- standard (designed),
- economic,
- ecological,
- appointed (prescriptive).

By the character of measures necessary for ESP implementation, three components: **technological, structural,** and **technical** are distinguished**.**

The **technological ESP component** considers an increase in the efficiency of production (extraction), preparation, transportation, and consumption of power resources and hence, a decrease in the energy intensity of products and services caused by the introduction of progressive energy efficient technologies.

The **structural ESP component** considers the reduction of relative contribution of energy intensive branches and manufactures to electrical engineering, manufacturing industry, and transport through the development of high-tech branches and production with low energy and material consumption.

The **technical ESP component** of the potential in rigorous calculations is determined based on the assumption that all out-of-date and inefficient equipment and technologies have been replaced by the most efficient ones. The expenses and restrictions on the replacement are disregarded.

1.2 Definition of the Energy Saving Potential

To determine (to estimate) the actual value of the energy saving potential, a reliable system is required for metering of and accounting for consumed energy resources. The list of entities subject to metering and accounting can change depending on the consumer type (thermal power stations, industrial enterprises, budgetary establishments, housing and communal services (HCS), etc.) and the employed technologies.

Generally the entities subject to metering and accounting are:

(1) fuel: gas, coal, and liquid fuel;
(2) energy: electricity, heat, and cold;
(3) energy carriers: hot water, tap water, wastewater, compressed air, ventilating emissions, etc.

All energy resource consumers should be provided by devices for metering the consumed energy and by devices fixing the basic operational parameters. Preliminary metrological inspection of their operating conditions is necessary to optimize the choice of the metering methods and means.

1.2.1 Metering of and Accounting for Energy and Energy Resources

The concept **metering** is very clearly defined. The simplest definitions are: (1) metering is a comparison with a measure and (2) metering is subject to metrological control and supervision and is possible where standards are strictly supported. The existing metering procedures establish all details of this procedure: types and characteristics of metering devices, required accuracy of metering of this or that parameter, and metering conditions.

At present the concept **accounting** has no officially accepted definition, but it implies a set of methods and procedures used to establish tariffs for the resources, to calculate their volumes that must be paid by the subscribers of energy saving organizations, and to calculate money arrived to the seller of the energy resource from the subscribers in accordance with the consumption norms and-or indications of the devices metering the amount of consumed resources and if necessary, of the devices registering the quality parameters of energy resources.

Accounting differs from metering by the fact that it deals with distribution of money and bills for payment of those or other resources based on the results of metering. If metering results are lacking, other coordinated approaches (algorithms) are used for the given system. Moreover, accounting should provide a decrease in payment for resources or services whose quality does not satisfy the conditions of the treatment on energy saving or of the agreement granting the utility services.

Since accounting and metering are different services, they are provided by different organizations: predominantly *billing companies* deal with accounting and *electricity metering companies* deal with metering.

The modern instrumental and computer base allows target-oriented energy monitoring to be organized at enterprises and considerable saving of energy and energy resources to be realized. (It is much more difficult to solve this problem for HCS.) For this purpose, the enterprise is supplied with the energy carrier metering equipment having outputs to obtain and process information: programmable logic controllers, operator workstations, and local computer networks. The input and output modules provide intellectual processing of discrete and analog signals and the local computer networks allow all information to be collected and visualized. The opportunity arises to process and to analyze every day the collected information, to establish deviations from standard power consumption, and to take necessary decisions of one of the three levels: operative control, administrative control, or strategic planning.

The operative control is usually the responsibility of the operating staff who compare actual energy consumption with the norm and undertake measures on non-admission of over-expenditure of energy carriers.

The administrative control is performed by the chief engineer of the enterprise/company and experts of structural divisions. They ensure assessment of works on energy utilization for a certain time period (typically for a week) and plan funding for maintenance and repair.

The strategic planning and monthly control are performed by the enterprise/company administration.

1.2.2 Organization of Energy Audit

Energy audit (EA) is the first and most important stage of elaboration and implementation of a complex of measures aimed at energy saving and improving the energy efficiency [4]. This term designates accumulation and processing of information on the utilization of energy resources to obtain trustworthy data on the amount of utilized energy resources and energy efficiency indicators and to reveal the possibilities of energy saving and improving energy efficiency with indication of the results obtained in the energy performance certificate. In addition to an analysis of the corresponding documents, the energy audit includes instrumental metering, balance tests of power plants, etc.

The concrete EA content is determined by its program specifying the scale, depth, and duration of the audit. The EA can be conducted with respect to a product, technological process, legal person, and private businessman.

An **energy audit** can be defined as: "The verification, monitoring and analysis of the use of energy and submission of technical report containing recommendations for improving energy efficiency with cost–benefit analysis and an action plan to reduce energy consumption."

It should be noted that the term "energy assessment" is sometimes used interchangeably with "energy audit" in some countries like the U.S.

1.2.3 Objectives and Types of Energy Audits

The objectives of an energy audit can vary from one plant to another. However, an energy audit is usually conducted to understand **how energy is used within the plant** and to find opportunities for improvement and **energy saving**. Sometimes, energy audits are conducted to evaluate the effectiveness of an energy efficiency project or program. The type of industrial energy audit conducted depends on the function, size, and type of the industry, the depth to which the audit is needed, and the potential and magnitude of energy savings and cost reduction desired.

An industrial energy audit can be classified into two types: a preliminary audit (walk-through audit) and a detailed audit (diagnostic audit).

In a **preliminary energy audit (Walk-through audit)**, readily-available data are mostly used for a simple analysis of energy use and performance of the plant. This type of audit **does not require a lot of measurement and data collection.** These audits take a relatively short time and the results are more general, providing common opportunities for energy efficiency. The economic analysis is typically limited to calculation of the simple payback period, or the time required paying back the initial capital investment through realized energy savings.

For **detailed (or diagnostic) energy audits**, more detailed data and information are required. Measurements and a data inventory are usually conducted and different energy systems (pump, fan, compressed air, steam, process heating, etc.) are assessed in detail. Hence, the time required for this type of audit is longer than that of preliminary audits.

The results of these audits are **more comprehensive and useful** since they give a more accurate picture of the energy performance of the plant and more specific recommendation for improvements.

The economic analysis conducted for the efficiency measures recommended **typically go beyond the simple payback period** and usually include the calculation of an internal rate of return (IRR), net present value (NPV), and often also life cycle cost (LCC).

The organizational and technological aspects of the EA in the developed countries are defined by legal national regulation acts being corrected as required.

The main EA aims are:

(1) obtaining objective data on the amount of utilized energy resources,
(2) defining the indicators of energy efficiency,

(3) defining the ES and IEE potential,
(4) elaborating the list of measures for ES and IEE and estimation of their costs. According to the agreement between the customer and the auditor, measures that differ from the typical ones can be included in the plan of measures on ES and IEE.

The main stages of the complex EA are: (a) preliminary analysis of the state of affairs with utilization of energy and energy resources, system of control over their flows at the enterprise/organization, (b) instrumental inspection, and (c) preparation of energy performance certificate and program (plan of measures) on ES and IEE.

The properly organized EA from measures on the estimation of the efficiency of energy and energy resource utilization and tests of the state of energy saving is transformed into an element of the system designed to ensure the stability of functioning of the enterprise/organization by increasing reliability and maintenance of resources on the required level as well as by strengthening of market positions and market expansion. For this purpose, the EA should be based on a complex energy economics analysis. In this case, examined entities are not only technical and economic indicators of the enterprises/organizations, but also investment, marketing, and other types of activity.

To this end, the energy audit itself should be as effective as possible, i.e., should satisfy a number of requirements, including

- integrated approach,
- reliability of results,
- informative results,
- feasibility.

Investigations and elaborations are being continued aimed at improving the efficiency of the energy audit, first of all, due to the integrated approach to its conduction implying inclusion of a larger number of components characterizing the energy efficiency, profitability, reliability, and quality of delivered electric energy. Great attention should be given to the improvement of the quality of information support and increase of its objectivity and productivity.

Recently the problem of improving the efficiency and attractiveness of the EA has been considered due to its cooperation and coordination with the ecological audit and the audit of economic and financial activities of consumers of energy and energy resources.

Based on the complex audit, programs for the strategic development of the enterprise/company are elaborated, including

- asset management strategy,
- equipment repair programs,
- technical and equipment modernization programs,
- investment management software,
- estimation of alternative business development strategies,

- profit maximization programs for holding subsidiaries and related business (synergetic effect) and algorithms for optimal profit redistribution (including holding asset management systems).

A set of requirements is also imposed on the audited entities:

- assistance to the personnel conducting the EA,
- provision of all necessary technical and technological documentation (executive energy communication schemes, data on fuel and energy consumption equipment, devices for metering fuel and energy resources (FER), regime maps, etc.) as well as documentation on financial and economic activity (branch and inter-branch standards and specifications, tariffs, consumption limits, agreements on FER delivery, warehouse fuel stocks accounting, data on FER consumption for own needs and transited and supplied to other consumers, data on losses, etc.),
- provision of statistical reports on product output and cost and amount of utilized FER.

Four EA types can be distinguished by the terms of work:

- pre-starting (pre-operational),
- primary,
- periodic,
- extraordinary.

Four EA types can also be distinguished by the scope of EA work:

- express inspection,
- full instrument inspection,
- integrated inspection,
- inspection of technological processes.

For each concrete EA type and audited entity it is necessary to adjust the indicators and their sets that would demonstrate results of the purposeful activity of the entity in the field of ES and IEEE and would be stable and reliable. The absolute values of the indicators allow the efficiency of energy consumption at enterprises, organizations, and establishments of one branch with analogous products to be compared. The analysis of temporal indicator dynamics for one and the same entity is no less important.

The EA has four main stages:

- *preliminary stage*: General acquaintance with the enterprise, its general plan, buildings and constructions, technological processes, and power supply systems;
- *documentary inspection (audit)*: Acquaintance with the available reports on planned and actual FER consumption, consumption standards adopted at the enterprise, and tests of energy equipment and their established periodicity. Treatments on the energy resource delivery, accounting, and invoices on payment are also studied if necessary together with their compliance with the treatment obligations on the supplied amounts, tariffs, and so on. In this stage, the situation with FER

consumption in the enterprise is analyzed and possible ways of obtaining additional more detailed information are outlined. Based on the obtained materials, the program of the third stage is planned;
- *instrumental inspection*: Obtaining information on energy supply and consumption that was not reflected in enterprise reports. The results obtained are compared with standard, certified, and calculated ones, and a conclusion is made about the effectiveness of the power supply system;
- *interpretation and analysis of the results obtained*: Based on the documentary and instrumental inspections, measures on ES and IEE are outlined and on the required amount of capital investments and terms of their recoupment are estimated.

In the stage of formulating recommendations, it is necessary:

- to formulate the technical essence of the suggested improvements and the principle of expenditure saving;
- to determine the components of the equipment required for the implementation of recommendations and their estimated cost, including delivery and commissioning;
- to consider all possible methods of decreasing expenses for promotion of the recommendations;
- to elucidate possible side effects of the recommendations influencing the actual economic efficiency;
- to estimate the total benefit from the suggested recommendations taking into account the foregoing.

The set of activities takes the form of an ES and IEE program with indication of terms of organizational and technical measures, funding sources, predicted saving in value and quantity terms, payback periods, responsible persons, and so on. In addition, by the EA termination, the executor prepares the following documentation:

- EA act (report) of the standard form,
- confirmed by FEB metering,
- energy performance certificate.

Recommendations on improving the FER utilization efficiency should not influence negatively the work of the audited entity, ecological characteristics of the working equipment and technological processes, level of safety and comfort of the working personnel, consumer product quality, etc.

The EA funding sources depend on the form of ownership of the audited entity. For example, the following order is established in Russia:

- FER consumers completely funded from the state budget pay for the EA from public funds,
- private/corporate entities pay from their internal funds,
- entities with mixed funding of their main activity from federal and local budgets, respectively, pay for the EA also from these two budgets. In all cases it is necessary to minimize the expenses to intensify the EA.

The energy audits of complex entities, for example, such as thermal power stations representing the system thermal energy source (TES/boiler-house)—system of heat transport—heat consumer are time consuming and expensive. One of the methods of reducing the EA time and cost is automation of the EA process in all its stages. The basis for the automation is automated collection and accumulation of indications from various counting devices widespread in power stations and electronic filling of typical report templates in the format of office packages. Practice has shown that this information is frequently not processed and not analyzed subsequently due to its huge volume that requires considerable human and material resources. This increases terms of the EA and makes it even more expensive.

In the first EA stage—collection of the initial data on the entity—filling of preliminary prepared questionnaires, the information for which was chosen from typical reports of an enterprise (for example, an electric power station) can be eliminated. Using modern software, databases can easily be transformed from the office packages into the questionnaires on the energy audit and thereby not only to save time, but also to exclude a number of possible errors caused by the human factor when transforming the data from one form to another.

In the stages of evaluation and implementation of the ES and IEE potential, mathematical modeling is rather promising. A set of mathematical models constantly updated and supplemented allows one to estimate comprehensively the audited system and to calculate various loading regimes within short time. In addition, a constantly increasing set of mathematical models allows one to save much time and many efforts during subsequent energy audits of entities with analogous equipment.

In the final EA stage, an energy performance certificate is developed or the already existing certificate is refined. Typical templates of the energy performance certificate and its content are approved by the corresponding documents and hence automated filling of the energy performance certificate becomes possible, because all initial data have already been obtained in the process of the EA.

The next level of the EA automation is the development of applied program complexes called *Automated Workplace of the Energy Auditor*. As applied to the EA of thermal power stations (TPS), the data contained in reports on fuel economy of power station equipment provide the initial information for the program complex. When conducting the EA of energy supply systems with several sources, the program complex allows the influence of each of them on the change in the specific fuel consumption required for generation of electric energy and heat to be estimated for the base period.

The EA practice shows that these complexes for automated information processing provide: (a) reduction of the operation time and material and human resources, (b) convenience of obtaining initial information (input data sheets of the program module correspond completely to the report templates), (c) possibility of output data application in various hierarchical levels of energy structures, (d) possibility of revealing indicators of energy efficiency from the results of calculations, etc.

1.3 Economic and Energy Efficiency and Financial Attractiveness of Energy Saving

1.3.1 Characteristic of Energy Saving Measures and Energy Saving Potential

The efficiency and attractiveness of projects on ES and IEE change in wide limits due to large differences in complexities of works (measures), costs of materials and accessories, personnel skills, and many other factors. The estimated efficiency of energy saving can be illustrated by Fig. 1.2.

Here three types of energy saving measures are shown:

(1) organizational measures (putting in elementary order energy utilization), providing up to 15% of saving potential; their cost lies within the limits up to 500 rub/TOE, and the investment payback period is about one year. They are called ***quick response measures***;

(2) Predominantly technological measures (up to 50% of the total potential) that cost from 500 to 1500 rub/TOE and have the investment payback period from 1 to 5 years. They are called ***base measures***;

The implementation of the expensive measures (projects) with long payback period is usually postponed to a later period and is included in plans of capital repair work, modernization, or replacement of the equipment.

(3) Investment measures (implementation of the remaining potential) with a payback period of about 5 years. They are called ***high cost, but highly effective measures***.

From Fig. 1.2 it can be seen that the higher the potential, the more expensive it is.

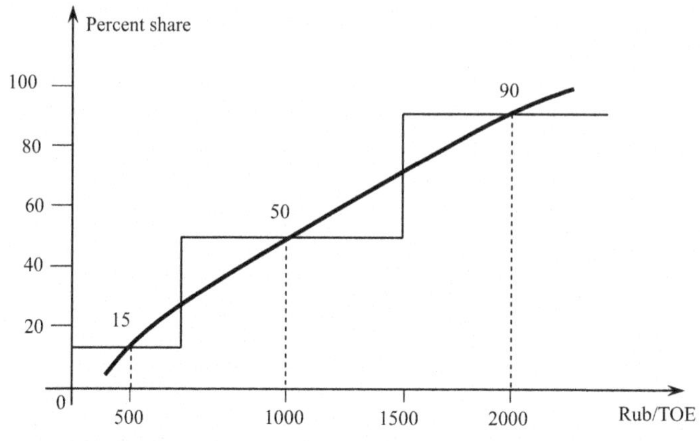

Fig. 1.2 Percent share of FER saving depending on its cost

The quick response measures can be developed and implemented within a year, and they have the significant effect and insignificant expenses. (Measures for increasing the level of awareness of the population in energy saving, stimulating and motivating the energy saving behavior of the personnel, granting the right to manage funds saved as a result of ES and IEE, specifying rules of purchasing the equipment for energy efficient technologies.)

The base measures are more radical; they promote rapid cost-effective and financially attractive investments. (Introduction of standards (specifications) on energy efficiency in such sectors, as HCS, industrial and public buildings, industrial equipment, funding of energy efficiency projects by banks and leasing companies, etc.)

High cost, but highly effective measures promote elimination of main causes for low energy efficiency; in most cases, they guarantee a most substantial saving of energy resources but require higher initial expenses. They involve large-scale organizational measures (on state, regional, or municipality levels): pricing reforms, modernization of electric energy and gas markets, transition to integrated scheduling of transport various types, deep HMS reforms, etc. as well as main technical/technological activities in manufacturing, building, transport, illumination, etc.

The energy saving is economically expedient only when it is cheaper than the replaced resource.

By the degree of coverage of various energy resources, these measures can be classified as follows:

- *Complex* measures that encompass at once all or some of the independent energy resources: fuel of different types, electric energy, heat, water, etc. Such measures are important for the implementation of the corresponding energy saving projects in regions, municipalities, etc. or in medium-sized enterprises;
- *closed* measures focus at once on all components of one of the independent directions of energy saving, including (a) production of electricity, heat, water, etc.; (b) their transportation to a consumer; and (c) utilization of energy carriers;
- *compound* measures include one or two rather than all above-mentioned measures (a, b, or c) of one of the independent directions of energy saving (they are important for projects of energy saving by different consumers and manufacturers);
- *simple* measures address only one component (a, b, or c) of one of the independent directions of energy saving (they are important for implementation of projects of energy saving by individual manufacturers of energy resources or HMS management companies).

When it is possible to apply simultaneously different **energy saving methods, it is expedient to subdivide them into the following types:**

- alternative methods that cannot be combined and only one of them must be chosen from the available list;
- consecutive methods that have no alternatives and can be used in combination with others.

Financial indicators of an investment project will depend on the sequence of a fulfillment of energy saving measures. As a whole, the indicators of the internal rate of return lower than for typical commercial projects are characteristic for investments in energy saving projects. This is due to a longer time of their implementation and a need for considerable investments.

To determine the optimal methods of investments in the ES and IEE projects, the *energy saving potential* of the investments is subdivided into **economically attractive, economically expedient,** and **financially attractive** (the implemented energy saving potentials (ESP) have the same names).

The economically attractive ESP is the part of the technical potential whose implementation is expedient according to the basic criteria of economic efficiency: discounting norms, alternative cost (for example, the export price of natural gas), ecological (decrease in emissions of CO_2 and other polluting substances), and other indirect effects and external factors. Consideration of these effects together with the increase of prices of the energy resources and the decrease of capital expenses cause the increase in the percentage of economically expedient investments. The investments into such projects are made by the government in interests of the country. They are not attractive for private investors.

The investments in the ES and IEE projects are considered economically expedient and financially attractive if the expenses on saving energy unity are less than the expenses on commissioning of the new generating facility. Such projects should guarantee the return to a society and-or private investors on the capital employed.

Economically expedient investments provide saving of the energy and money resources for the state during implementation of the investment project.

Financially attractive ESP is a part of the economic potential, which is expedient for implementation within the limits of existing market conditions, prices, and restrictions. For private investors, in particular, for households (individual investors), higher discounting rates are established. This is caused by three reasons: (a) owners of households do not take risks associated with investments in an improvement of energy efficiency; (b) they should take loans under higher interest rates for investments; (c) they have other, more attractive spheres for the investment of available assets. Financially attractive projects provide recovery of invested funds to private investors.

From all possible measures on ES and IEE, only an insignificant part is economically expedient and financially attractive. Their difference is caused by (a) different discount rates for the state and private investments; (b) manifestation of indirect effects from energy saving; (c) allowance for the positive and negative external factors which requires additional expenses for implementation of ES and IEE projects. Among the most significant factors is the reduction of CO_2 emissions being the consequence of many projects on increasing energy efficiency.

The economically expedient and financially attractive measures are selected from technically effective ones by a comparison of expenses and benefits from their implementation.

Some of the investment energy saving projects appear too expensive or provide too insignificant saving during their implementation that makes them unattractive for energy consumers.

1.3.2 Estimation of Economic Consistency of Investments in Energy Saving Projects

The choice of energy saving measures should be preceded by the technical and economic analysis of their efficiency and determination of their payback period and the sequence of their implementation. In this case, the ***economic analysis*** is understood as a process of evaluation of the economic efficiency of the project for the society or saving as a whole irrespective of the beneficiary from the project implementation, and the ***financial analysis*** is understood as a process of estimating the project implemented by an individual business entity to evaluate the financial return on the invested capital. The economic efficiency reflects results of implementing energy saving measures and is determined by the difference between money incomes and expenses on the implementation of these measures as well as reflects the change in the demand for fuel and energy resources as a result of replacement of expensive fuel by a cheaper one.

The economic efficiency of the ES and IEE measures can be estimated for a typical investment project whose key parameters are:

(1) Current net cost,
(2) return on investments,
(3) Net profit/internal rate of return,
(4) Payback period of capital investments, etc.

A complete list of indicators estimating the efficiency of implementation of measures comprises approximately 15 items. The four above-indicated key parameters of the economic efficiency of the ES and IEE measures are briefly considered below.

Parameter 1. Current net cost as a basic characteristic of these estimations includes: (a) incremental capital expenses for implementation of the project and b) all additional expenses or effects from the implementation of the measure. They include expenses on elaborating a business plan or a feasibility study of the measure, design and exploration work, design and survey work, purchase of the basic and supplemental equipment, and installation and commissioning work. The incremental capital expenses on energy saving measures (E_{IC}) are calculated as a difference between the expenses for new equipment that will allow specific energy consumption to be reduced to the level of *practical minimum* or *average international indicator* and the expenses on installation of the new equipment with energy efficiency comparable to that of the equipment installed at present in this country.

The capital expenses for these projects are estimated with allowance for the coefficient K_{IC} that transforms the total capital investments into the equivalent annual expenses. The K_{IC} value depends on the possible return on the invested capital for alternative investments (on the discounting rate).

Additional expenses or benefits K_{EN} can include the change of the operating expenses within a year avoiding the need for capital investments, or external effects relevant to the concrete energy saving project. When calculating C_{ES}, benefits are reflected in the form of negative expenses K_{EN}. The total annual expenses on the project of improving the energy efficiency ($K_{IC}\, E_{IC} + K_{EN}$) are expressed in terms of the energy saved (for example, in kW h of electric energy or in m^3 of a gas). For this purpose, the total annual expenses on the project are divided into the total annual saving obtained as a result of project implementation.

A comparison of C_{ES} with average prices of energy carriers allows one to answer the question: *Is the project interesting from the viewpoint of the economic or financial efficiency?* The energy saving measures can be considered efficient if:

- current net cost is positive (and the higher, the better),
- payback period is short,
- profitability index ≈30%,
- internal norm (rate) of profitability is greater by 10 points than the profitability index of the project.

State investors (including regional administrations and municipalities), non-commercial, and public organizations can be interested in investment in projects with low profitability if these projects benefit the society as a whole or the given territorially-administrative units.

Parameters 2 and 3. Return on investments and internal rate of return in the course of technical and economic analysis can be estimated either through the specific energy consumption per unit of production/energy intensity of the product or through the economic efficiency (E, rub/yr).

Absolute economy E_m (the efficiency of energy saving measure in natural units: TOE, kW·h, or GJ) can be represented as

$$E_m = (W_1 - W_2)P_2 \tag{1.1}$$

Then the ***economic efficiency*** (the efficiency expressed in rubles) can be determined from Eq. (1.1):

$$E = E_m C = (W_1 - W_2)P_2 P. \tag{1.2}$$

P in Eq. (1.3) is the price (tariff) of the energy resource expressed in rub/TOE, rub/GJ, rub/kW·h, etc.; E_m is the effect in material (energy) form; W_1 and W_2 are power consumptions (FER) in TOE or kW·h per unit of product before and after implementation of the measure (specific consumption), the change of which is caused by the given measure; and P_2 is the annual output of the product after implementation of the measure expressed in unit of product.

The *relative economy*, as a rule, is expressed in percentage and is determined as the ratio of the absolute economy (in rubles) to the initial expenses (also in rubles):

$$E_{\text{rel}} = \frac{E_{\text{m}}P}{M_{\text{m}}} = \frac{(W_1 - W_2)\text{P}_2\text{P}}{M_{\text{m}}}, \tag{1.3}$$

where: M_{m} denotes money spent (planned) for the implementation of the measure, in rubles. Nowadays the indicator E_{rel} is usually determined by the management of the corresponding organization.

It is accepted to characterize the efficiency of operation of a separate aggregate/ installation (turbogenerator, boiler, etc.) by the ratio of the useful output power of the installation to the expanded power.

Parameter 4. Payback period (PBP) of capital investments is the period of time from the launch of the project to the moment of time when the income from its implementation will cover its initial cost.

Two methods are used to estimate it:

- the payback period is defined as expenses on the implementation of the energy saving project divided into the annual economic benefit of its implementation—a simple payback period (the simplified method);
- The time period is determined during which the economic effects from money put in a bank at some interest and from money investment in the suggested energy saving project become equal (or the period of repayment of the bank loan taken for project implementation is determined).

The payback periods calculated by these methods practically coincide if the period calculated by the first method is within 2–2.5 years.

The given criterion allows one to estimate the time period during which the planned reduction of energy resource consumption, losses in grids, reduction of emissions of greenhouse gases, etc. can be attained. In energy saving the PBP is the indicator that determines the time period required for compensation of investments at the expense of the net financial income (benefit) from energy saving. The payback period is called the least time period during which the net profit becomes and further remains non-negative. In the estimation of the efficiency it, as a rule, acts only as a restriction.

In other words, the PBP is the time period during which the total capital and operational expenses become comparable with the total benefit:

$$\text{PBP} = \frac{I_0}{T_e \Delta E - C}. \tag{1.4}$$

Though the PBP allows one to estimate it simply enough for capital investments in the project, in some cases it yields incorrect results.

The time moment when the net income becomes equal to zero corresponds to the PBP taking into account discounting (discounting takes into account depreciation and reduction of expenses at different points in time to the time of launch of the examined project). In the simplest case the efficiency of the energy saving project can be estimated from the simple payback period of the investments necessary for the implementation of these projects:

$$T_{PB} = \Sigma I/\Sigma E \text{ year,} \tag{1.5}$$

where: ΣI denote total investments in the energy a saving project, and ΣE denotes the annual economic benefit from the project implementation including saving of energy resources and other expenses of the enterprise associated with the implementation of the proposal minus annual expenses on the implementation of these measures.

It is obvious that the consideration of discounting leads to an increase in the PBP; this, in turn, leads to the fact that orientation on the given indicator causes discrimination of projects with extended PBP. A maximal value of the PBP can be estimated if we know the following quantities: the coefficient α characterizing a number of annual payments for a hypothetical loan borrowed from a bank at the annual rate equal to the discount rate (depends on the amount of the loan), and d is the discount rate:

$$\alpha < \text{PBP} < \frac{1}{\alpha} < \frac{1}{d}. \tag{1.6}$$

Since energy saving measures can be compared for rather long periods, changes in prices for energy carriers can be considerable. This factor should be considered in calculations. In this case, as additional indicators we can use are:

- change in time of the price for energy carriers;
- Profitability index (PI) equal to the ratio of the reduced benefits to the reduced capital investments. The simple ratio of all reduced benefits to all reduced expenses disregarding signs of these quantities is conventionally used. The ratio greater than unity demonstrates the efficiency of energy saving measures. The projects characterized by a higher PI value are more efficient;
- The criterion of minimal expenses is usually used when it is difficult to estimate benefits from the introduction of the measure. In this case, the expenses calculated using various methods are compared and a more favorable variant is chosen (the least expenses provide the maximum results). When choosing a measure among the competing measures, in addition to the criterion of the least expenses, additional information on the feasibility of the project is used.

When calculating the efficiency of the project with allowance for discounting, the payback period is determined by the term of repayment of the loan taken for its implementation. The loan is taken at an interest considering also expenses for project insurance. The payback period is the term of repayment of the loan from the economic benefit that takes into account all expenses. The projects with long PBP (exceeding 2–2.5 years) are considerably influenced by the interest rate of the bank loan.

An example of the schedule of repayment of the loan taken for the implementation of an energy saving project at an interest and the economic benefit directed on repayment of the loan is shown in Fig. 1.3.

Investments in energy saving projects are economically attractive to municipal entities when the cost of energy saved as a result of project implementation is less

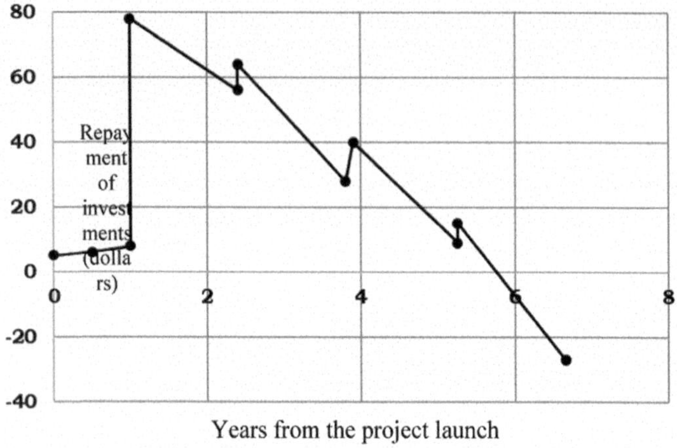

Fig. 1.3 Estimated payback period of the loan taken on the implementation of an energy saving project

than the cost of energy that must be purchased to ensure comfortable vital activity of the population, entrepreneurship entities, and social sphere.

Expenses associated with its consumption according to the valid tariff. There are various obstacles for the consumers that hinder the implementation of this indicator of rationality. A serious obstacle is lacking investments from energy consumers for implementation of high cost, but highly efficient measures (Fig. 1.3). The cost of the saved energy, in this case, starts to exceed the existing tariff on the saved energy [5].

A more exact indicator of economic attractiveness of the project is the ***reduced net income*** (RNI). It represents the sum of all annual discounted financial benefits B_n and expenses for project implementation. This is the most generalized criterion showing incomes for the entire period of project implementation. The RNI calculated during the entire period of project implementation with the chosen discount rate gives the integrated effect from its implementation. The RNI is calculated as the difference between the sum of the income from the energy saving measures (benefit) reduced to the project launch time and the sum of capital expenses for the project made by this time:

$$\mathrm{RNI} = -I_0 + \sum_{i=0}^{N} \frac{B_i - C_i}{(1+d)^i}, \tag{1.7}$$

where: B_i is the annual financial benefit from energy saving equivalent to the amount of saved energy carrier in the i-th year of project implementation multiplied by the energy tariff:

$$B_i = T_e \Delta E.$$

For constant B_i and C_i values we have

$$\mathrm{RNI} = -I_o + (T_e \Delta E - C)/\alpha. \tag{1.8}$$

The RNI can be considered as the current income or benefit from the investments. To calculate the RNI, the corresponding discount rate should be determined, discounting of benefits and expenses should be considered, and then the reduced costs should be summarized. If the sum is positive, the net discounted income will also be positive, and the measure can be recommended for funding. The project can be considered economically viable if the discounted benefits exceed the discounted expenses. It is obvious that when choosing between mutually exclusive projects, preference should be given to the project with higher (positive) RNI. The higher the RNI, the more economically attractive is the project. The energy saving project with negative RNI value can also be implemented when a significant environmental effect is possible. Then other funding sources must be tapped, including budget ones.

To compare energy saving projects of different scales, the indicator of internal profitability norm (IPN)—the maximum discounting rate *(d) at* which the net benefit is not lower than the net expenses—is typically used. The IPN equals to a maximum percentage on the loan which can be repaid from incomes obtained during project implementation so that it remains self-benefiting, that is, can generate funds whose amount is sufficient to cover completely the capital and operational expenses.

Whereas the calculation of the RNI of a power saving project yields the absolute efficiency at a preset discount norm, the IPN of the measure shows its relative efficiency and then is compared with the required income norm on the invested capital. As the IPN can determine the priorities that differ from the RNI criterion, the measures are ranked using simultaneously these two main criteria.

The amount of energy saved as a result of investments in the energy efficiency includes both direct and indirect savings. The implementation of projects aimed at improving the energy efficiency leads to the direct reduction of the final energy consumption, whereas the consumption of primary energy is reduced indirectly due to the reduction of expenses for energy transformation into the form acceptable for the end users.

The cost of energy is not only its social price, since a high social price has to be paid for the adverse environmental impacts of energy production. Generation of electric energy (like many other types of economic activities) causes social expenses that basically do not refer to electric energy generation. Elimination or control of harmful effects causes additional expenses of electric energy generation.

The scales of environmental effect can be expressed in terms of the environmental performance index of an economic (industrial) system, and the value of this index is considered as an integral component of the decrease in the expenses of primary energy carriers for the examined period. In this case, it becomes possible to estimate a change in the amount of polluting substances taking into account specific indicators of their emissions into the atmosphere, hydrosphere and lithosphere.

The integrated approach to the estimation of the efficiency of energy saving measures usually includes the following target indicators:

- generalized target indicators,
- target indicators reflecting saving of individual FER types,
- target indicators of energy efficiency in housing fund,
- target indicators of energy efficiency in systems of municipal infrastructure.

References

1. Ushakov VYa, Kharlov NN, Chubik PS (2017) Energy saving in the enterprises of fuel and energy complex: textbook. TPU Publishing House, Tomsk Polytechnic University Tomsk, p 176
2. Sumper A (2012) Electrical energy efficiency. Technologies and applications/Wiley/A. Sumper, A. Baggini, 1th edn, p 434
3. Fokin VM (2006) Fundamentals of energy saving and energy audit. M.: Publisher "Maschinostroenie - 1", p 256 (in Russian)
4. Hasanbeigi A (2010) Industrial energy audit guidebook: guidelines for conducting an energy audit industrial facilities/A. Hasanbeigi, L. Prise. LBNL-3991 E
5. Accelerating Energy Efficiency in Small and medium-sized Enterprises (SMEs) (2015) Powering SMEs to catalyse economic growth. http:www.iea.org/publications /free publications/SME

Chapter 2
Methods of Estimation Electrical Power Consumption by Industrial Enterprises

This chapter deals with existing approaches and methods for managing electrical power consumption (EPC) by industrial enterprises. When considering these approaches, the focus is primarily on modeling, rationing, and forecasting. Methods for managing (EPC) by industrial enterprises are analyzed through both direct and indirect methods. The chapter describes measures for implementing these methods depending on the production profile of the industrial enterprise.

2.1 Possible Approaches to Managing Electrical Power Consumption

Improving the efficiency of (EPC) by industrial enterprises, as with entities in all sectors of the economy in developed countries, is one of the most important tasks.

An effective way to enhance the competitiveness of industrial products is to reduce the cost of consumed power, that is, to decrease its specific energy intensity. Achieving this goal requires improving the management system of EPC. Traditional methods of EPC management used in industrial enterprises today include modeling, rationing, and forecasting. Unfortunately, these methods do not sufficiently consider the radical changes in the economies of some countries due to their transition to a market economy. This leads to significant errors in modeling, rationing, and forecasting EPC. The transition to a market economy enhances the importance of considering the peculiarities of technological processes of industrial enterprises to improve the efficiency of their EPC.

To date, many diversified industrial enterprises have been studied in terms of energy consumption efficiency. Enterprises operating on a continuous mode (with a continuous nature of production) deserve to be a specific object of such studies. This is based on at least two circumstances:

I. U. Rakhmonov et al., *Electric Consumption by Industrial Enterprises*, Power Systems, https://doi.org/10.1007/978-3-031-62676-0_2

- Most industrial enterprises operating in a continuous production cycle (such as metallurgy, chemistry, oil refining, etc.) are energy-intensive. As a result, they achieve significant economic effects from measures that ensure a reduction in specific energy intensity, even by a few percentage points.
- In the patterns of energy consumption by enterprises with a continuous production cycle, belonging to various economic sectors, there are many common features. This allows for the development of mathematical models and programs acceptable for such enterprises.

Existing methods for assessing EPC primarily consider hourly productivity as the main influencing factor. Particularly, depending on the technological process, calculations for workshops are made with and without considering technological factors. Energy characteristics have been established for them, and the patterns of changes in EPC have been identified. EPC unit of production for different types of products is grouped, and it is within these groups that EPC norms are established [1].

In estimation EPC, it is more essential to identify the patterns of EPC by consumers. Graphically, the dependencies of EPC parameters (required power, total EPC) on changes in the technological process are shown. However, these energy indicators are used as intermediate characteristics. The EPC charts for producing finished products are constructed without considering the constant consumption of power for auxiliary but inevitable purposes (such as lighting, heating, pumping and compressor operations, etc.); it is recommended to use these characteristics when analyzing the power consumed in production. That is, they relate only to the main technological equipment and do not reflect energy consumption for auxiliary purposes.

The progressive method of assessing specific electrical power consumption SEPC allows for the use of corresponding norms of specific expenditures for forecasting energy consumption and determining energy load in planning and designing. Here, the SEPC for production processes in workshops and enterprises is an important technical and economic indicator of the energy management of industrial enterprises [2]. However, to date, the calculation and analysis of SEPC at enterprises do not meet modern requirements, as they do not take into account several factors:

(a) The energy intensity of the range of products manufactured and raw materials received at different stages of the technological process;
(b) Changes in the quantitative ratio of the product range of the enterprise, technological regimes, and structural features of the units of this industrial sector.

Currently, more than 70 types of finished products are manufactured. Moreover, the same type of finished product is produced using different technological schemes with varying SEPC, depending on the composition of the technological units and the quality of the processed raw materials.

Scientifically substantiated norms of SEPC allow for highly accurate forecasting of EPC at the level of the workshop, unit, and enterprise, as well as controlling the effective and targeted use of power, identifying points of low energy efficiency for cases when the volume of manufactured products does not correspond to the normative amount of consumed power.

The system of rationing of EPC is a factor determining the policy and efficiency of energy saving and it is based on the results of research and developments by a number of scientists and specialists.

Usually, in the practice of producing final products at an enterprise, specific norms of consumption of various energy resources (fuel, thermal and electrical energy) are developed. The importance of these norms lies in that they allow for determining the SEPC for a specific period.

When utilizing the experimental method for rationing SEPC, it is necessary to measure power expenditures for each production process as stipulated by the regulations. It should be noted that determining this value requires significant labor efforts, as it involves the use of a large number of measuring instruments. The application of this method in industrial enterprises is complicated by the extensive amount of electrical equipment [2].

The calculation-analytical method for determining SEPC norms involves calculating them based on the passport data of the used electrical equipment. This calculation takes into account factors like load, operating modes, and other individual factors that affect the SEPC. The presence of numerous energy expenditure components makes this method of element-by-element calculation extremely labor-intensive. The calculation-statistical method for rationing the SEPC is based on processing statistical data on absolute and SEPC over a given period; all factors influencing the SEPC are considered in establishing norms. These norms are set based on data from EPC measuring devices, taking into account the resulting values of the volume and nomenclature of the produced goods [2].

The development of substantiated forecasts for EPC norms involves several stages [3, 4]:

- comprehensive technological and energy research;
- In-depth analysis of electric balances;
- Critical use of reporting and statistical materials with the application of mathematical-statistical methods.

The absence of methods for estimation the influence of technological factors on electrical loads and EPC does not allow for effectively estimation for unforeseen and planned deviations due to technological conditions, including reserves for energy saving and predictive evaluation of the results of production process improvements.

The analysis of electrical loads and EPC by industrial enterprises highlights the need to identify the influence of technological factors of EPC on its individual parameters. This will increase the reliability of the analysis and calculations of EPC for individual production units and the entire industry.

Methods of operational planning and management of power consumption by industrial enterprises are used in assessing the error of calculating norms of EPC by existing methods, which take into account the statistical dependence of SEPC on the volume of produced final products [5–7].

In practice, a method for rationing SEPC for industrial enterprises is used, allowing the determination of norm values by accounting for the dependence of EPC on the volumes of produced goods, as well as the specific consumption determined by the mathematical processing of actual indicators of SEPC [8, 9].

A methodology for rationing and analyzing SEPC has been developed using probability theory and methods of mathematical statistics. In this methodology, based on a small amount of primary data, only a limited number of factors influencing EPC (production efficiency, installed capacity) can be considered.

The application of a systems approach to determine the norms of absolute and SEPC overall reveals the scientific and methodological aspects of the problem. The use of a large number of uncertain parameters as primary data, the high level of flexibility, and the possibility of modeling nonlinear processes enhance the practical value of these research data [5].

The classic method is often used for rationing the consumption of power by industrial enterprises. This method involves determining SEPC indicators at the level of technological equipment, workshops, and the enterprise as a whole. SEPC in [6] are determined as the volume of EPC relative to the manufactured product, that is, calculations are based on single-factor analysis. Accounting for the specifics of the technological process, a multitude of parameters influencing the norms of EPC, and the possibility of obtaining generalized indicators are the advantages of this method.

It is known that for determining normative indicators with high accuracy, it is advisable to conduct power measurements consumed by each device. In this case, the application of the method for determining the SEPC [10] is appropriate.

In the operation of the energy sector, primary information is of great importance, which is used in planning normal operating modes. Based on the conducted forecast, the following tasks are solved:

- Determining effective methods and approaches for forecasting EPC;
- Calculating real and optimal operating modes of energy systems;
- Assessing parameters of reliability, quality of electrical energy, etc.

Well-founded forecasting of EPC allows industrial enterprises to avoid penalties.

Of the approximately 150 existing forecasting methods, 20–30 main ones are used in practice. They are classified according to the following characteristics: formalization degree, principle of operation, and methods of obtaining forecasted information.

For solving the tasks of forecasting EPC, firstly, a deep analysis of the EPC of individual production sections is conducted, and then algorithms for predicting EPC as a whole for production are used.

It is known that forecasting methods are divided into formalized, heuristic, and complex. Each type of forecast is characterized by it is own advantages and limitations. When using formalized methods, only quantitative indicators are obtained. The development of these forecasts assumes that in the future, the system will develop as it did in the past. These methods include extrapolation and regression methods, the method of group accounting of arguments, etc. [11].

The accuracy of heuristic methods depends on the professional level of the specialist, who, through their knowledge and practical experience, can anticipate qualitative changes in the development of the forecasted object. These methods are used in cases where there are abrupt changes in the development of the system and are divided into individual methods and collective expert assessment methods [11].

Integrating formalized and heuristic forecasting methods into a unified system (comprehensive forecasting) enhances the reliability of the forecasts. The existence of over a hundred such methods, as previously noted, underscores the critical importance of selecting the right method, which depends on the nature of the forecasted object and the level of knowledge and skills of the relevant specialists [12].

The method of determining predictable indicators of EPC based on time series analysis provides high accuracy in establishing predictable indicators for group consumers but does not allow for accurately predicting indicators for individual consumers. To overcome this limitation, it is necessary to supplement this method by considering the influence of internal production factors such as ambient temperature, the proportion of cloudy days, the length of working days, days of the week, and the patterns of switching on and off energy-intensive technological equipment [13, 14].

In estimation EPC, determining the degree of impact of technological factors on it is the main cause of overestimating the assessed expenditures. In the process of estimation, each factor is considered separately. This approach enables the creation of highly accurate models for forecasting the energy consumption of industrial enterprises. However, it is important to note that in some cases, considering many factors when determining forecasting values can lead to a significant increase in errors.

Forecasting the volumes of SEPC can be performed at different levels: throughout the enterprise, for workshop or by technology. In forecasting the volumes of SEPC, empirical dependencies of consumption on the quantity of products are used. Unfortunately, this method has not been tested on real data regarding the production of main types of products and EPC; it also does not allow for accounting for many parameters that affect the SEPC.

Given the complexity of calculating the volumes of consumption and production of power, its quality, and several other indicators necessary for developing a forecasting model, a method for optimizing the EPC of energy-intensive units has been proposed to improve the accuracy of calculations.

In forecasting EPC using the principal component method, the coefficients of the regression model are calculated based on solving quadratic programming tasks, which is a major advantage of such energy consumption forecasting. To ensure the regulatory accuracy of the forecasting, the optimization of the main criteria parameters is carried out using the genetic algorithm method [3, 15, 16].

Analysis of forecasting models of electrical loads of industrial enterprises showed that highly accurate predictable indicators can be obtained using artificial neural networks. This approach is noted for its advantages in forecasting under conditions of uncertainty [17–19].

Multi-stage methods of planning and managing EPC by industrial enterprises address issues of assessing power expenditure by accounting for factors that influence the inaccuracies in determining EPC indicators and ways to reduce them [20].

Based on systemic methods of managing the consumption of energy resources by industrial enterprises, issues of improving methods for rationing and forecasting EPC are considered. This has made it possible to implement the developed models and algorithms for managing EPC at several enterprises [21].

Automated systems for regulating energy consumption by industrial enterprises enable solving tasks of forecasting EPC, taking into account the nature of the technological process, the composition of power-consuming equipment, and its operating mode. Interesting results have been obtained using artificial neural networks, genetic algorithms, and several other methods [22].

To estimate the consumption of electricity, it is extremely important to take into account theoretical and practical aspects of rationing and forecasting EPC by industrial enterprises, as well as the possibility of optimizing technological processes and mathematical modeling. Executing these stages provides an opportunity to enhance production efficiency and reduce EPC.

In [23], modeling EPC by industrial enterprises includes the following stages:

(1) Development of EPC model for energy-intensive technological equipment (using probabilistic methods, which are used in the analysis and planning of their EPC);
(2) Identification of factors influencing EPC and quantitative estimation of their impact;
(3) Analysis of the peculiarities of EPC by enterprises in the coal industry;
(4) Creation of a mathematical model of EPC in individual links of the technological process based on the theory of Markov processes.

As noted earlier, the power supply system of an energy-intensive industrial enterprise represents a complex system consisting of several subsystems, and each major power consumer can be considered as a separate object of EPC regulation. In this regard, in [24], the assessment of EPC analyzed the energy consumption of each type of equipment and identified laws of their changes; the influence of seasonal factors on EPC was analyzed. Considering these issues during the assessment of EPC enables the smoothing of load graphs for these consumers.

In the field of modeling, rationing, and forecasting EPC at industrial enterprises, comprehensive methods have been developed that encompass all economic sectors, including the management of EPC in metallurgy, textiles, oil and gas, oil refining, machinery, construction, and other sectors of heavy and light industry. These proposed methods are based on determining the ESPC for each type of product. This allows for the selection of optimal operating modes for technological equipment, the development of measures for the rational use of various types of energy, and the identification and assessment of power saving reserves.

The analysis of existing approaches to managing EPC by industrial enterprises shows that much has been accomplished in this direction, particularly in assessing power expenditure through modeling, rationing, and forecasting EPC. Considering that technological processes in industrial enterprises are being rapidly improved today, with new energy-efficient and energy-saving advanced production technologies being introduced, these factors must be taken into account in developing methods for assessing EPC costs.

2.2 Implemented Methods for Managing Electrical Power Consumption

Analysis of the current state of this issue shows [25, 26] that there are several types of industrial enterprises today, differing in the type of EPC management, which depends on the nature of technological processes. Two main groups of power consumption management methods can be identified—direct and indirect. In the first group of methods, the object of management is the flow of electrical energy coming into the enterprise, whereas, in the group of indirect management methods, the object of management is the schedule of electrical load (consumption of active power).

The second method mentioned involves smoothing the load curve by controlling the duration of operation and downtime of technological equipment without affecting the continuity of the technological process.

Figure 2.1 illustrates the scheme of direct and indirect control of EPC and the main directions of its implementation. It shows that industrial enterprises primarily enhance the efficiency of power use through indirect control.

In the method of indirect control, three subgroups can be identified [27–29]:

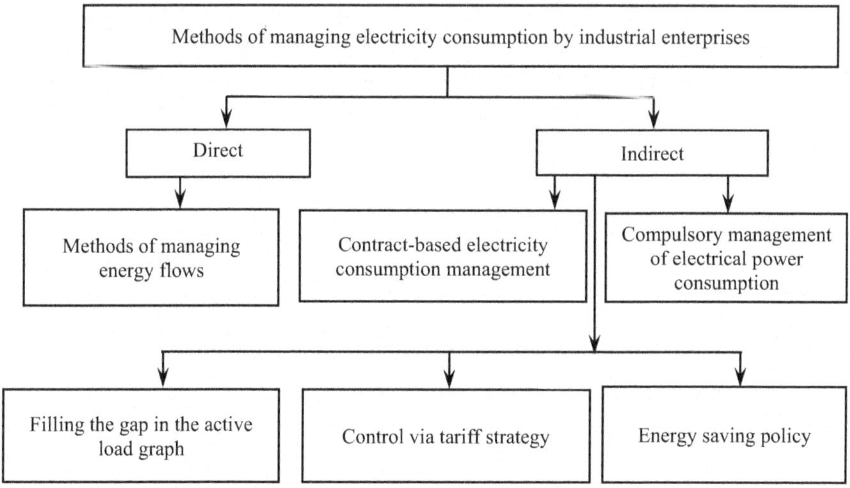

Fig. 2.1 Methods of managing electrical power consumption by industrial enterprises

- Energy consumption management based on a contract between the power supply company and the consumer;
- Compulsory management of EPC;
- Management of the power load mode.

It is noted that in industrial enterprises with continuous operation cycles, 80% of the technological (main) equipment operates continuously; auxiliary equipment consumes only 20% of the energy used for the full technological process. Therefore, any method of managing EPC should ensure the uninterrupted continuity of the technological process.

As noted above, the indirect control method is the primary approach used in industrial enterprises. Below are briefly described measures for implementing this method, depending on the industrial enterprise's production profile.

Contract-based electricity consumption management. Issues of EPC management are addressed by applying the "consumer-regulator" principle and identifying consumers willing to work according to this principle.

Compulsory management of EPC. This involves limiting the consumption of active power during peak load in the power system, that is to say, system-wide optimization of the technological process without affecting the main technological process and shifting the load from the time of maximum EPC to the time of minimum consumption, thus smoothing the power load graph. This is a classic method of managing EPC. Based on this, the reduction of the maximum load and, consequently, the increase in the time of minimum load are achieved, and the uneven distribution of load over time is also reduced [30, 31].

Filling the gap in the active load graph. If smoothing the power load graph is the first classic method of managing EPC, then filling the chart is the second classic method. These methods provided technical and economic efficiency before the introduction of modern methods of EPC management by industrial enterprises. They are widely used even today. It is known that depending on the operating mode of electrical equipment at industrial enterprises, the graph of electrical loads by active power varies within a wide range. Reducing this range—filling the chart of electrical loads, is carried out by connecting (commissioning) additional power-consuming equipment [29].

The energy saving policy is implemented not only through increasing energy efficiency but also through energy substitution expanding the use of thermal, solar, wind, geothermal, and other alternative energy sources. The implementation of these programs to achieve high technical and economic indicators is an important task in managing EPC [30].

Control* via *tariff strategy. This method of indirect management is one of the classic methods of management, the application of which has allowed for achieving high energy efficiency at several industrial enterprises in countries with developed economies.

In Fig. 2.2, measures for implementing this method are presented. The measures carried out within the framework of these methods can be conditionally divided into

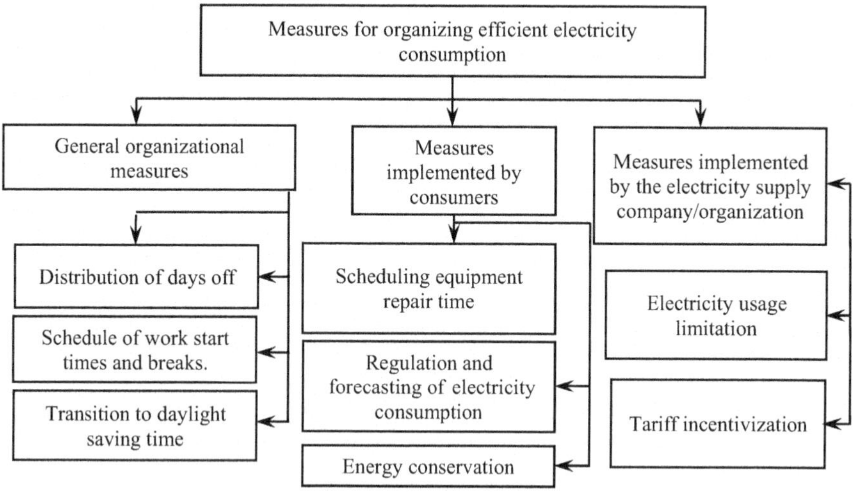

Fig. 2.2 Implementation of methods for managing electrical power consumption by industrial enterprises

the following three groups: (1) general organizational measures; (2) measures carried out by power consumers; (3) measures carried out by energy supply organizations.

Below is a brief analysis of the measures included in each group.

General organizational measures. This category includes measures related to the proper distribution of working hours at an industrial enterprise: rest for workers, start and end of the working day, breaks between working hours, and transfer to summer working time. The determining factor here is the human factor. When implementing them, it is necessary to study the opinions of the employees of the enterprise, conduct relevant surveys among them, and determine the above time intervals based on the results of the survey.

Measures implemented by consumers. As shown in Fig. 2.1, this set of measures includes traditional power management activities such as scheduling repair time for the installation, rationing and forecasting EPC, and implementing energy-saving measures. When drawing up a schedule for implementing these measures, continuous technological processes must be ensured.

Below we will take a closer look at the second and third measures.

Measures implemented by the electricity supply company/organization. These measures include limiting (setting a limit on) EPC and stimulating through tariff policy. Both measures allow for managing electrical power consumption at the enterprise level. Based on the analysis of the daily load schedule of the industrial enterprise, EPC can be managed using a tariff policy. In this case, while ensuring the continuity of the main technological process, it is recommended to shift the load to the time of minimal power cost—the nighttime period. Of course, such a shift is carried out in accordance with the daily load schedule of the area where the manufacturing enterprise is located. In addition, managing EPC modes is divided into two groups: passive and active management methods.

Passive management method. This method involves periodic disconnection of technological electrical equipment operating on the "consumer-regulator" principle in accordance with the recommendations of the power supply organization. The passive management method implies the analysis of the technological process, studying the operating modes of the equipment, researching the control of the EPC by equipment operating in the mode of repetitive and short-term work, and determining the permissible frequency of their disconnection without a negative impact on the technological process.

Active management method. In this method, the mode, duration of operation, and downtime of electrical devices operating on the "consumer-regulator" principle are determined based on the study of the load schedule for active power. In this case, as mentioned above, the main criterion is ensuring the continuity of the technological process.

Based on the existing methods and approaches to EPC management discussed above, a comprehensive method for managing EPC at industrial enterprises with continuous production is proposed, see Fig. 2.3.

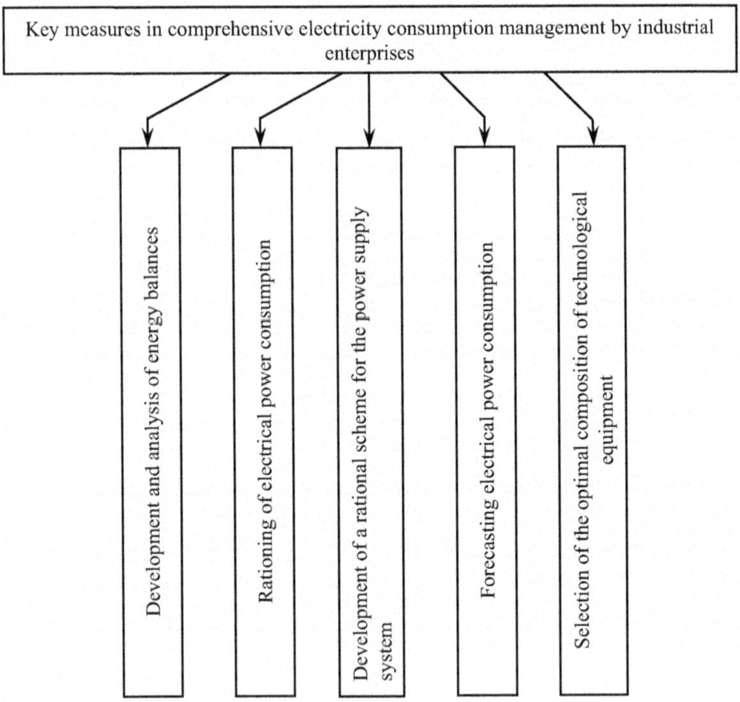

Fig. 2.3 Key measures in comprehensive electricity consumption management by industrial enterprises

Rationing of EPC. In industrial enterprises, the rationing of EPC is one of the classic methods of managing this process. The determination of precise and scientifically justified norms of SEPC allows for the assessment of EPC efficiency and the identification of reserves for power savings. The normative values of SEPC, in turn, are used to determine the forecast of the required power generation depending on the volumes of products manufactured by industrial enterprises [32].

Development of a rational scheme for the power supply system. The issue of rational choice of power supply schemes for an industrial enterprise plays a vital role in the comprehensive provision of processes for rational reception, conversion to standard voltages, and distribution of power to consumers.

Forecasting EPC. One of the main directions in managing EPC at industrial enterprises is to improve the accuracy of forecasting EPC limits. In general, this method of managing EPC is one of the classic methods, which over time is being improved with the use of artificial intelligence methods [33].

Selection of the optimal composition of technological equipment. This activity is carried out with an in-depth analysis of the technological process of the industrial enterprise. In doing so, it is necessary to choose such an optimal composition of equipment that in this set the criteria of optimization are met: minimal EPC and maximum volume of manufactured products.

Thus, from the above, it can be concluded that in all methods of managing EPC, the main requirement remains ensuring the uninterrupted technological process at the enterprises. Failure to meet this requirement leads to significant technical and economic losses.

References

1. Kiyko S (2021) Formation of a portfolio of energy saving projects at metallurgical enterprise. Innov Mater Technol Metall Mech Eng 12(3):71–81
2. Borys S, Iryna K, Dmytro S (2023) Increasing the reliability of mechanisms of metallurgical equipment that uses SHS resource-saving technology. Techn Sci Technol 2(32):78–82. https://doi.org/10.25140/2411-5363-2023-2(32)78-82
3. Anna B, Olesia Y, Dima S, Anton R, Elina T (2021) Neural network forecasting of energy consumption of a metallurgical enterprise. Innov Technol Sci Solutions Ind 2(1):14–22. https://doi.org/10.30837/ITSSI.2021.15.014
4. Komarov S, Shakhverdov S (2023) Management foundations of resource saving at modern industrial enterprises of the extractive industry. State Municipal Manage Scholar Notes 1(3):23–32. https://doi.org/10.22394/2079-1690-2023-1-3-23-32
5. Shenets Ya (2022) Evaluation of the energy efficiency of industrial furnaces based on the modeling of fuel consumption modes. Energy. In: Proceedings of CIS higher education institutions and power engineering associations, vol 65, issue 2, pp 169–180. https://doi.org/10.21122/1029-7448-2022-65-2-169-180
6. Kai L, Hou M, King PS, Han W, Tan (2021) Industrial electrical energy consumption forecasting by using temporal convolutional neural networks. In: MATEC web of conferences, vol 335, p 02003. https://doi.org/10.1051/matecconf/202133502003
7. Larisa V, Elena M, Igor M (2023) Digital management of resource efficiency of fuel and energy companies in a circular economy. Energies 16:3498. https://doi.org/10.3390/en16083498

8. Yusufjan S (2023) "Green" energy: experience of foreign countries, problems and prospects 3(2):247–253
9. Karnaukhov BN (2018) Application of energy characteristics of units for consumption norming and electrical power consumption analysis. Norming of electrical power consumption and energy balances of enterprises
10. Koptsev LA, Koptsev AL (2019) Consumption rationing and forecasting of electrical power consumption in industrial enterprises. Ind Energy 2:18–23
11. Grinev AB, Novikova OV (2014) Improvement of energy consumption management at industrial enterprises. In: Proceedings of the international conference efficient energy, pp 27–28
12. Palchikov AS (2018) Existing methods of forecasting electrical power consumption for objects in the metallurgical industry. Modern scientific research and innovation, issue 9 [Electronic Resource]
13. Bazhinov AN, Ershov EV (2011) Decision trees in the problem of selecting significant factors for forecasting electrical power consumption in the metallurgical production. Bull Cherepovets State Univ 4(3):10–13 (in Russian)
14. Bazhinov AN, Ershov EV (2011) Electrical power consumption forecasting as a means to improve the efficiency of metallurgical production. Metallurg 11:34–37 (in Russan)
15. Tikhonov EE (2018) Forecasting methods in market conditions: a tutorial. Nevinnomyssk, p 221
16. Kirpichnikova IM, Solomokho KL (2014) Study of electrical power consumption forecasting methods for a sales enterprise. Electrical Syst Complexes 3(24):39–43 (in Russian)
17. Toropov AS, Tulikov AN (2018) Hourly electrical power consumption forecasting of a regional power system using artificial neural networks. Bull Irkutsk State Techn Univ 5:143–152. https://doi.org/10.21285/1814-3520-2017-5-143-152
18. Sharifullin VN, Sharifullina AV (2019) Neural network forecasting of electrical power consumption by an industrial enterprise. Bull Kazan State Energy Univ 13(4):6–12 (in Russian)
19. Delegodina LA (2018) Energy consumption forecasting in an automated control system for unified energy systems using artificial neural networks. In: Computational and network resources, pp 66–72
20. Koptsev LA, Zhuravlev YP (2008) Energy efficiency management of an electric steel melting shop. Vestnik SUSU 17:13–20 (in Russian)
21. Xinxiang W (2021) Application of computer control system in electrical automation. J Phys Conf Ser 1982:012127. https://doi.org/10.1088/1742-6596/1982/1/012127
22. Rakhmonov IU, Niyozov NN, Kurbonov NN, Umarov BS (2023) Forecasting of electrical power consumption by industrial enterprises with a continuous nature of production. In: E3S web of conferences, vol 384, p 01030, 1–4. https://doi.org/10.1051/e3sconf/202338401030
23. John S (2019) Energy consumption analysis and modeling in coal mines: a case study of the ruhr area, Germany. Dissertation for the Degree of Doctor of Engineering, Technical University of Munich
24. Thomas B (2017) Assessing the efficiency of electric power usage for technological purposes in steel production facilities. J Modern Manage Technol 5(77):124–128. ISSN 1234-5678
25. Robert M (2015) Approaches to industrial energy management using process energy characteristics. Doctoral Thesis, Technical University of Berlin
26. Fisher Emily A (2019) Analysis of mathematical models for forecasting econometric time series. Br Technol J 7(2):61–73
27. Pierre D, Lefevre M (2018) Methods and tools for managing energy supply and consumption. Educational Manual for Students in "Electrical Supply" Specialization, École Polytechnique, Paris, p 194
28. Rakhmonov IU, Nematov LA, Niyozov NN, Reymov KM, Yuldoshev TM (2020) Electrical power consumption management from the positions of the general system theory. J Phys Conf Ser 022054. https://doi.org/10.1088/1742-6596/1515/2/022054
29. Alexander S (2023) Management of strategic planning based on merger and acquisition deals and its impact on the efficiency of the Russian electric power complex. Econ Manage 17:141–150. https://doi.org/10.14529/em230313

30. Hlushchenko YaI, Chernenko NO, Korohodova OO, Moiseienko TIE (2022) Industry 4.0 technologies to ensure energy saving in the utility sector. In: Review of transport economics and management, pp 117–138
31. Rakhmonov IU, Najimova AM (2022) Forecast of electrical consumption in cotton-textile cluster enterprises taking into account the volume of production. AIP Conf Proc 2647:030011. https://doi.org/10.1063/5.0104791
32. Li Y (2023) Research on difficulties and countermeasures of electrical power consumption inspection in power enterprises. Appl Sci Innov Res 7(4):76–82. https://doi.org/10.22158/asir.v7n4p76
33. Rakhmonov IU, Reymov KM, Dustova SD (2020) Improvements in industrial energy rationing methods. J IOP Conf Ser MIP: Eng 862:062070. https://doi.org/10.1088/1757-899X/862/2/062070

Chapter 3
Energy Balances

In the chapter, the structure, purpose, types, and application areas of energy balances (EB) are discussed, along with the technology of their composition. EBs are classified according to the types of energy, time and stage of development, purpose, stages of the life cycle of energy resources, qualitative characteristics, and the scale of the object. Described are **actual, normalized, rational, optimal, and ideal EBs**. Examples of EBs of an enterprise and its subdivisions are provided.

3.1 Energy Balances: Purpose, Types, and Applications

An EB is a system of indicators that characterizes the process of energy transformation or supply to consumers, reflecting the equality of the supplied energy on one hand, and the sum of useful energy and losses on the other. A comprehensive characteristic of the volumes of extraction, processing, transport, transformation, and distribution of primary, processed, and transformed types of fuel and energy, starting from the extraction of fuel-energy resources (FER) and ending with their delivery to energy-consuming installations, is represented by the fuel and energy balance (FEB), which for brevity is usually called an energy balance (EB). (In this monography, the authors use this particular term.) The compilation of an EB is necessary for determining the need for fuel and energy and for planning the development of an enterprise, region, city, industry, or country [1–3].

The (EB) of an FER consumer, as the basis for forming economic, business, production, informational, and other management mechanisms in the energy sector, is widely used. With sufficient completeness and reliability, an EB allows for the optimization of the structure of production of energy resources and energy, their distribution, transmission, and consumption, types of energy-generating and energy-consuming installations, and types of energy carriers for various technological processes and conditions.

I. U. Rakhmonov et al., *Electric Consumption by Industrial Enterprises*, Power Systems, https://doi.org/10.1007/978-3-031-62676-0_3

EB are classified according to seven main criteria [3–5]:

1. ***By types of energy***: electrical, thermal, fuel.

For the majority of consumers, the main energy sources are heat and electricity, so in many cases, **partial EBs**—thermal and electrical—are sufficient. At industrial enterprises consuming large volumes of local energy resources—compressed air, gases, cold, water, etc., partial EB are compiled for the corresponding energy carriers.

This classification group also includes a consolidated EB, in which all types of energy are converted into one energy units and summed up. In Russia and in most former union republics of the USSR, such a unit is a ton of fuel equivalent (toe) with a calorific value of 7000 kcal/kg, in Europe and the USA—a ton of oil equivalent (toe) with a calorific value 12,000 kcal/kg.

2. ***By time and stage of development***: reporting, current, prospective and project.
3. ***By purpose***: for the needs of the main production, for the needs of auxiliary production, for municipal and domestic needs.
4. ***By stages of the life cycle of energy resources***: during the extraction of primary energy resources, during their processing and enrichment, transportation, conversion into thermal and electrical energy, during consumption and use.
5. ***Based on qualitative characteristics***: actual, normalized, rational, optimal, ideal.
    - *Actual balances* characterize actual energy consumption. These balances are compiled on the basis of real-time instrument accounting.
    - ***By time and stage of development***: reporting, current, prospective and project.
    - *Rational* ones are compiled to improve and rationalize energy use.
    - *Optimal* ones provide for complete technical re-equipment using the most advanced energy-saving equipment.
    - *An ideal energy balance* is one compiled only for useful energy, without taking into account losses and exclusively for research purposes.
6. ***By the scale of the object***: state, region, municipality, enterprise, division (workshop), energy-intensive installation (unit).
7. ***According to the parameters included in the balance sheet base***: direct and reverse balances. Both balances allow us to judge the efficiency of using energy resources and energy.

In the direct balance, the basic parameters are the energy supplied to the consumer and the usefully utilized energy. (Compiled in cases where direct measurements can be conducted.)

The final useful use of energy by various consumers can be determined based on the following considerations [6]:

- In lighting systems, usefully utilized energy is determined by the level of illumination on the working surface;
- In transport systems, by the amount of transport work performed;
- In power processes, by the amount of mechanical energy of the final actuator;

- In thermal processes, by the theoretical energy consumption for heating, melting, evaporation, and conducting endothermic reactions;
- In heating, ventilation, and hot water supply, by the amount of heat actually supplied to the consumer;
- In electrochemical processes, by the energy consumption theoretically necessary for the process;
- In the processes of storage, processing, conversion, and transport of fuel and energy, by the quantity of energy carrier obtained after the corresponding conversion.

In *the reverse balance*, the basic parameter is the energy loss at the consumer's site (at the facility). Several categories of losses are distinguished:

- Related to the implementation of a given technological process and caused by technical imperfection of the equipment;
- Economically justified during the technological process at the traditional level using conventional equipment;
- Economically justified during the technological process at an advanced scientific and technical level, ensured by modernization and reconstruction of equipment and technological schemes and the introduction of new technologies;
- Related to the deregulation of the technological scheme, operating equipment in suboptimal modes;
- Associated with the use of low-quality energy resources;
- Caused by the need to maintain an elevated level of reliability, safety, durability, etc.;
- Associated with providing increased comfort, specific working conditions, information, and festive events.

All losses of energy and energy resources can be grouped into 3 unequal and vaguely defined groups, caused by the following reasons:

(a) Imperfection of operating conditions of equipment, requiring organizational and regulatory measures for its elimination;
(b) Imperfection of the technological process, requiring technical and technological restructuring and rationalization of existing equipment for its elimination;
(c) Imperfection of technological schemes and equipment, requiring radical reconstruction of production, replacement of technology, and large investments for elimination.

The main principles and tasks of compiling the EB are as follows [7, 8]:

- The balance is formed primarily on the basis of existing forms of state statistical reporting;
- The comparison of the arrival, expenditure, and losses of energy resources is carried out at each stage of production, transmission, and final use of energy resources;
- Estimation energy losses at each stage allows for identifying ways to reduce them;

- Measuring the flows of energy and energy resources, planning, and analysis allow for preparing management decisions and justifying tariffs for energy and energy services;
- Improving the structure of the balance is aimed at increasing the efficiency of energy resource use, replacing inefficient resources and technologies with more efficient ones and/or reducing the cost of using energy resources;
- Estimation energy and energy resource losses at each stage of their use for decisions on replacing inefficient resources and technologies with more efficient ones;
- Estimation the energy-saving potential.

As a result, the energy audit and compilation of the EB, the difference between the actual consumption of energy resources and energy and progressive standards of their consumption is established. This difference is the **energy-saving potential (ESP)**.

Establishing progressive norms for the consumption of fuel-energy resources (FER) by their various consumers is a complex task, mainly solved during balance tests. Due to many years of using state, industry, and enterprise balances, its compilation technology is well developed.

Depending on the purpose, EB can be classified based on the following characteristics: development time, energy consumption object, intended use, the totality of analyzed energy flow types, development method, and compilation form.

Based on the time of development, EB are divided into:

(1) Project-based, compiled during the development of the corresponding project.
(2) Planned, compiled for the upcoming planned period considering tasks to reduce energy consumption norms.
(3) Reporting (actual), compiled based on reported (actual) data for the past period.
(4) Promising, compiled for the forecasted period considering fundamental changes in technology, production organization, and the enterprise's energy management.

EB are compiled for objects of energy consumption such as enterprises, productions, workshops, sections, units, installations, etc.

Depending on the intended use, FEB are developed for technological processes, heating and ventilation, lighting, etc.

Based on the totality of analyzed energy flow types, EB include:

- Specific EB for individual types and parameters of consumed energy carriers.
- Consolidated EB for the total consumption of energy resources and the direction of their use.

In terms of development method, EB are divided into:

- Experimental, compiled based on actual measurements of parameters and energy flow consumption.
- Calculative, compiled based on the calculated energy consumption of the production under consideration.

- Experimental-calculative, compiled using both actual measurements and calculations.

Currently, three forms of energy accounting are used [9]:

(1) Using measuring instruments;
(2) By calculation;
(3) Experimental-calculative method.

Each accounting form involves the following procedures:

- Recording primary indicators of the quantity and quality of all types of energy, both received externally and consumed at the enterprise, as well as generated and delivered externally (if the enterprise engages in such activities);
- Operational accounting of energy consumption using specialized instruments;
- Adjusting the calculated parameters of energy carriers based on instrument readings;
- Determining energy consumption by calculation for those departments and production areas where there are no accounting instruments.

Primary documentation and operational accounting of primary energy carrier indicators, as well as primary load accounting, are performed based on measuring instrument readings.

Primary energy accounting documentation includes daily operation reports of units, operational (watch) logs, load charts, meter readings, etc. Primary documentation should contain primary indicators characterizing the technical condition and quality of equipment maintenance. All primary documentation indicators are recorded in daily reports with intervals of no more than 0.5–1 h.

Secondary documentation consists of reports on the operation of installations and the energy sector, worksheets (workbooks) reflecting final and average equipment performance indicators for a shift and a day. They form the basis for compiling monthly energy balances, quarterly technical operation reports, and summarizing final indicators.

The most accurate method is determining parameters using appropriate control and measuring instruments (*instrumental determination*). Other methods are auxiliary and are mainly used to distribute energy consumption among consumers with a common account.

The calculative method of energy and energy resource accounting provides only approximate information and is used when installing instruments is impractical.

The experimental-calculative accounting method is recommended when there are no energy carrier accounting instruments on the site, and the calculative method does not provide the necessary accuracy. It is based on the combination of one-time measurements of energy indicators by control and measuring instruments and their subsequent determination by calculation.

In industrial enterprises, depending on their purpose, various types of EB are used. Their classification is shown in Table 3.1 [10].

Table 3.1 Classification of EBs

Classification characteristics of balances	Types of balances
By stages and time of development	Project-based, planned, actual, prospective, prediction
By objects of development	Balances of technological operations, machines, units, production lines, workshops, enterprises, industries, countries
By intended purpose	Balances for individual types of technological processes, auxiliary processes
By the number of types of fuel, energy, and energy carriers covered by the balance	For individual types of fuel and energy; for all types of energy resources used at the enterprise
By methods of obtaining initial information for compiling balances	Experimental, calculative, experimental-calculative
Depending on the level of energy consumption	Actual, normative, optimal
By form of presentation	By production-territorial feature, balances of machines, sections, workshops, etc., analytical balances, synthetic balances, internal circulation of FER
By methods of evaluating the work performed using the energy resource	Energetic, exergetic

Based on the nature of the decomposition of the expenditure part, energy balances are divided into "consumption balances" and "utilization balances" of fuel-energy resources.

The former shows the distribution of energy among consumers. These types of balances help identify the most energy-intensive consumers and establish the structure of energy resource consumption at the enterprise.

The second type involves dividing the energy supplied to the consumer into the useful component of expenditure and losses by elements.

The variety of balance types entails differences in the methods of their compilation. The choice of method depends on the purpose of the balance, the required degree of detail, the method of obtaining initial information, and its available volume. As noted above, there are three main methods for obtaining data to compile a balance: experimental (instrumental), calculative, and experimental-calculative.

The most comprehensive assessment of the thermodynamic efficiency of energy-consuming equipment can be given using ***exergy balances***. To date, significant experience has been accumulated in using exergy for analyzing energy consumption processes. The exergy assessment of energy processes is conditioned by the peculiarities of converting one type of energy into another. Depending on this, the theory of exergy divides all types of energy into those that can be fully (without restrictions) converted into any other types of energy and those that convert with an increase in entropy.

The main indicators of energy consumption efficiency are:

- Efficiency coefficient of energy installations;
- Coefficient of useful energy usage;
- Coefficient of useful usage of energy by its types and parameters;
- Specific energy carrier consumption (consumption per unit of produced product).

When analyzing an enterprise's energy consumption, it is compared with similar data from similar enterprises, including generalized indicators of the state and development of the energy economy, such as: electrification coefficient, thermal-electric coefficient, electric-fuel coefficient, electric intensity of products, and thermal intensity of products. Analysis of EB helps solve tasks related to issues of replacing one energy carrier with another in technological processes and installations, using secondary energy resources as a substitute energy carrier available at and outside the enterprise, choosing directions for fuel and energy saving, and developing specific measures for fuel and energy saving [11].

The development and analysis of energy balances are aimed at solving the following main tasks:

- Estimation the actual state of energy use at the enterprise, identifying the volumes of energy resource losses and their nature;
- Developing a plan of measures aimed at reducing energy resource losses;
- Identifying and estimation the reserves for saving fuel and energy;
- Improvement of rationing and development of scientifically substantiated norms of fuel and electricity consumption for manufacturing products;
- Determining rational levels of energy consumption in production processes and installations;
- Defining requirements for the organization and improvement of accounting and control of energy carrier consumption;
- Obtaining initial information for solving issues of creating new equipment and improving technological processes to reduce energy costs, optimizing the structure of the enterprise's EB by choosing optimal directions, methods, and sizes of using supplied and secondary energy resources, improving internal production economic calculation and the system of incentives for energy resource saving.

3.2 The Structure of Energy Balances

The structure of the energy balance of a company can be represented by the following data for the purpose of compilation and analysis [12]:

- The volume and range of products manufactured, their cost including the energy component, etc.;
- Schemes of material and energy flows;
- A list and characteristics of the main energy-consuming equipment;
- Consumption of energy carriers;

- Work on the rational use of energy carriers in the enterprise.

The EB consists of *incoming* and *outgoing* parts. The incoming part reflects the list and amount of energy received in the form of various energy carriers: primary (fossil fuels and nuclear fuel) and secondary (electricity, gas, steam, water, air). The outgoing part reflects the consumption of energy carriers and energy in all their diversity, their losses during the transformation from one form to another, during transportation, and the energy stored in special devices (for example, pumped-storage power plants, batteries and accumulators, pneumatic, inertial energy storage devices, etc.).

When compiling the EB, all types of energy and energy carriers consumed at the enterprise, all directions and purposes of energy consumption at each site of the enterprise, as well as its losses, are considered.

Data on actual energy consumption at specific sites of the enterprise and by specific types of energy-consuming equipment (engines, electrical equipment, lighting devices, etc.) are obtained using meters for electricity, gas, steam, water, etc.

Depending on the type and quantity of energy carriers, the EB can be either specific, compiled only for one energy carrier, or consolidated, compiled based on the total volume of consumption of all energy resources used in the enterprise (in this case, the unit of measurement is tons of equivalent fuel). For compiling specific energy balances, the unit of measurement is the joule (J) and its derivatives (MJ, GJ), as well as kilowatt-hours (kWh).

EBs are compiled for two main types of energy consumed by industry, urban and rural economies—thermal and electrical energy.

In operational enterprises, EBs are compiled for individual units or their groups, workshops, and the enterprise. Based on the EB, an objective judgment is made about the efficiency of the use of electrical energy at the enterprise and in its divisions or energy-intensive units. The identified potential for reducing unproductive EPC to allow planning measures for energy saving and efficient energy use.

The most important results of compiling normalized EBs are:

(a) the possibility of rationing electrical power consumption in the main technological processes;
(b) determination of their energy efficiency and identification of ways to improve it. The balance of active energy (normalized EB) is the main type of balance, reflecting the actual mode of EPC and the level of its useful use.

The incoming and outgoing parts of the EB are based on the readings of active energy meters and calculated power. The expenditure part of the active electrical energy balance includes the following expenditure items:

(1) Direct electricity costs for the main technological process (for the manufacturing of products) with the allocation of useful EPC by energy-intensive equipment (in electric furnaces, compressor and pumping installations, etc.);
(2) Indirect electricity costs for the main technological process, due to its imperfection or violations of technological norms;
(3) Electricity costs for auxiliary needs (ventilation of workshop premises, workshop transport, lighting);

(4) Losses of electrical in power supply networks;
(5) The sale of energy carriers to third-party consumers related to social infrastructure objects (catering establishments, clubs, stores).

The final stage of the analysis of the actual balance of the enterprise is the compilation of a free standardized EB—the basis for assessing the reserves of electricity savings.

The reserves for saving electricity are divided into current ΔW_T, implemented with small expenditures in the current period, and prospective ΔW_n, the implementation of which is possible in the more distant future (3–5 years and more) through the implementation of measures requiring significant additional expenditures.

Current reserves for energy savings are determined by comparing the actual EB of an object with its EB, compiled based on technically justified individual losses using the formula:

$$\Delta W_c = \sum (\Delta W_{bi} - \Delta W_{ai}),$$

where ΔW_{bi} and ΔW_{ai}—are the losses of electrical energy in each object before and after the implementation of measures, respectively.

Prospective reserves for savings (ΔW_p) are determined by comparing two normalized energy balances—technically W_t and economically W_{eco} justified (prospective):

$$\Delta W_p = \sum \Delta W_{ti} - \sum \Delta W_{eco}.$$

Depending on the purpose, the following indicators are considered in the EB:

- The calculation period (reporting balances based on actual data for the past period, planned for the nearest planned period with tasks to reduce energy costs, design, compiled during the design of the object, etc.);
- The link in the life cycle of the energy flow (production, transformation, distribution, final use of energy resources);
- The type of energy carrier considered when compiling: (a) specific EBs—for individual types of consumed energy carriers, (b) consolidated EBs—for the total energy consumption.

The scheme of material and energy flows is accompanied by a description of the types and parameters of energy carriers, the state of use of secondary energy resources, and the system of accounting and control of energy and energy carrier consumption.

When evaluating the efficiency of FER use at an enterprise, the amount of ΔW_T as both used and supplied energy is summed up at all installations for all types of energy carriers used.

3.3 Electrical Balance of the Enterprise and Its Divisions

The EB is compiled using measurable parameters for which accuracy standards are established, and measurement methods are developed, standardized, and certified. The development and analysis are carried out by departments defined in the relevant industry standards and enterprise standards, in collaboration with design, research institutes, and specialized organizations.

1. At the stage of enterprise design (by the institute—the general designer), both specific and consolidated project EBs are developed for all objects of the enterprise. An analytical form of the balance is provided for the main energy-intensive installations.
2. At an industrial enterprise, EBs are compiled for use in the following time intervals:
   - For the nearest planned period (year, five-year plan)—a planned synthetic EBs (both specific and consolidated) for energy consumption objects and with a breakdown by purpose;
   - After the end of the reporting period based on the internal factory reporting documents—a reporting (actual) synthetic EB;
   - Once every five years (on the third year)—an actual analytical EB with a certain degree of detail for objects, purposes, types of energy carriers, etc.

For most industrial enterprises, the primary energy carrier is electricity, and therefore the EB is the main component of the overall technical energy balance. It is compiled for the enterprise as a whole, as well as for workshops, individual units, and their groups. The data from the EB allow for assessing the quality and efficiency of electricity use at the enterprise and its divisions, and for planning and implementing measures to improve the EPC.

In Table 3.2, the results of the technical and economic evaluation of the EF of FER are presented, using the example of a typical industrial enterprise. These results demonstrate SEPC the potential for energy savings for all energy-intensive processes. The normalized EB allows for standardizing the electricity costs for the main technological processes and for the manufacturing of finished products. The rationing of the balance is based on the consideration of individual EBs of the most energy-intensive units and workshops. Scientifically justified technological and workshop-specific norms of EPC are the basis for the normal consumption of electricity.

The rationing of losses in equipment is carried out based on its standard characteristics, and for the most energy-intensive units—based on the results of special control tests and calculations. Based on the main purposes of the EB, its primary objects are considered to be active power and energy. The balance of reactive power and energy is compiled taking into account the generation of reactive energy by compensating devices and is used when solving problems of compensating reactive power and voltage regulation.

Depending on the specific purposes and conditions of measurement, EBs by energy are related to a certain period: the duration of one working shift, one day,

Table 3.2 Assessment of technical and economic values of efficiency factor of fuel and energy resources (EF of FERs) by industrial enterprises

Types of processes	EF of FERs values								Ratio of existing to prospective
	Existing				Prospective				
	Generation and transformation	Transport and transmission	Consumption	Total	Generation and transformation	Transport and transmission	Consumption	Total	
Internal combustion engines	87	93	20	16	90	95	25	21	0.72
Electric drive	29	85	65	15	40	90	85	30	0.5
Lighting	29	85	20	5	29	85	40	10	0.5
Centralized heating supply	78	92	85	60	85	96	95	77	0.78
Decentralized heating supply	55	93	90	46	75	97	95	69	0.68
High-temperature fuel installations	87	90	35	27	90	92	65	54	0.5

a month, or a year. Depending on the degree of detail, *differentiated* and *structural electric balances* are distinguished. In the expenditure part of the differentiated balance, the costs of electricity for the main technological process and all types of accompanying losses are reflected, and in the structural balance, the electricity costs for the main process by types of technological equipment (machine, power, electrotechnology), as well as for ventilation, lighting, etc., taking into account losses (without dividing them into constant, load, etc.).

The incoming part of the EB of a unit/workshop includes the electricity received at the terminals of the electric receiver/inputs to the workshop, and for the entire enterprise—the electricity received from the power system and produced by its own sources.

The incoming part of the reactive EB of the workshop/enterprise also contains data on its generation by all sources of reactive power: capacitor banks, overexcited synchronous motors, and synchronous compensators.

The expenditure part of the EB contains the following main items:

(1) Direct energy costs by units for carrying out the technological process with the allocation of constant and load losses in electrical and technological equipment;
(2) Energy losses in intra-workshop (distribution) and inter-workshop (power) lines, general factory and workshop transformers;
(3) Other types of losses (startup, thermal, etc.), energy consumption by lifting-transport equipment, for ventilation and lighting;
(4) Energy consumption by auxiliary equipment and devices;
(5) Energy consumption by general plant consumers not related to the technological process (so-called subtenants, which include canteens, clubs, and other cultural and domestic structures);
(6) The sale of electricity to third-party consumers as part of its resale.

It is obvious, the number of items is maximal in the EB of the entire enterprise and minimal in the balance of individual units of technological installations. For the latter, a differentiated EB is compiled, and for the workshop—both differentiated and structural.

The income part of the workshop balance is formed by summing up similar items from the expenditure part of the EBs of the feeding feeders and is presented in the form of tables. An example of the income part of the daily EB of a mechanical workshop, which receives power from the factory's 0.4 kV network through three power and one lighting feeders, is shown in Table 3.3.

Reactive power is included in the shop balance because part of it is generated by compensating devices installed in the workshops and by synchronous motors operating with leading current.

Examples of the expenditure part of the differentiated and structural balances of the same workshops are given in Tables 3.4 and 3.5, respectively.

The electric balance of a plant/enterprise is formed by summing the balances of the workshops, the EPC of the general plant services and departments, the supply of

Table 3.3 The income part of the workshop EB

Article of income part of EB	Electrical energy	
Received from the plant network via feeders	P (kWh)	Q (kVARh)
No. 1	1810	1940
No. 2	1560	1450
No. 3	1590	1845
No. 4	270	–
Generated in the workshop: by capacitor banks, synchronous motors	– –	1840 –
Total	5230	3395

Table 3.4 Expenditure part of the workshop's differentiated EB

Article of the expenditure part of the EB	Electrical power consumption (kWh)					
	Feeder No. 1	Feeder No. 2	Feeder No. 3	Feeder No. 4	Total for the workshop	
					kWh	%
For the main technological process	940	845	815	–	2600	49.7
Permanent losses	556	520	456	–	1532	29.4
Load losses	128	92	90	17	327	6.3
Thermal losses	–	12	32	–	44	0.8
Startup losses	15	–	20	–	35	0.7
Losses in the workshop network	16	13	14	5	48	0.9
Lighting	–	–	–	248	248	4.7
Auxiliary needs	125	78	131	–	334	6.4
Household needs	30	–	32	–	62	1.1
Total	1810	1560	1590	270	5230	100

electric power to external customers, losses of energy in the main step-down substation, and in the distribution networks. A graphical representation of the consumption part of the EB is shown in Fig. 3.1 [13].

The passage describes the complexity of reflecting all consumers in a company's balance sheet, leading to a statistical discrepancy of up to 10%, with energy intake being higher than expenditure.

It then details the components of energy resource consumption for gas transportation, including:

- Compressing of gas;
- Automatic air cooling of gas;

Table 3.5 Expenditure part of the workshop's structural EB

Type of equipment and expense items	Electrical power consumption	
	kWh	%
Power equipment	3409	65.2
Electrical equipment[a]	1078	20.7
Handling equipment	209	4.0
Ventilation	125	2.4
Lighting	248	4.7
Household needs	62	1.1
Losses in the workshop network	48	0.9
Unaccounted for equipment	51	1.0
Total	5230	100

[a] Electric furnaces, heaters, welding equipment, etc.

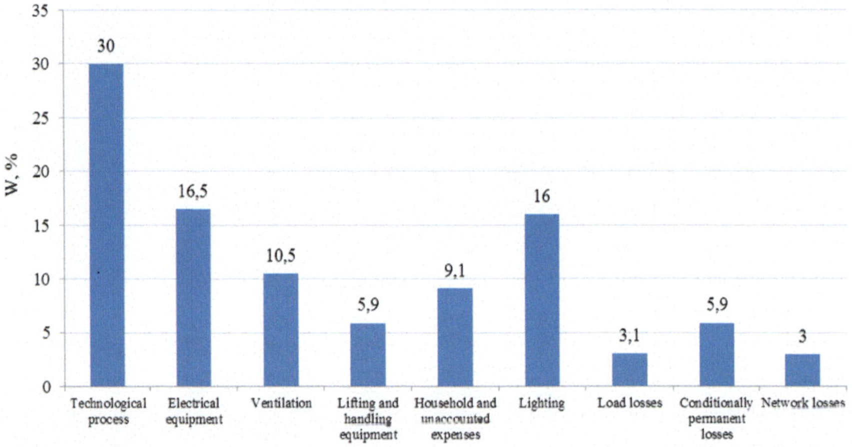

Fig. 3.1 Diagram of the EB of the enterprise

- Gas pumping;
- Losses of electrical energy in the transmitting and transforming elements of the power supply system.

The standard consumption of gas during its transportation includes:

- Gas consumption by compressor stations for technological needs.
- Technological losses of gas at compressor stations.
- Technological losses of gas in the linear part of the main gas pipeline.

The EB of a company for an energy audit reveals the difference between the quantities of supplied energy and the energy efficiently used (see Fig. 3.2) [14].

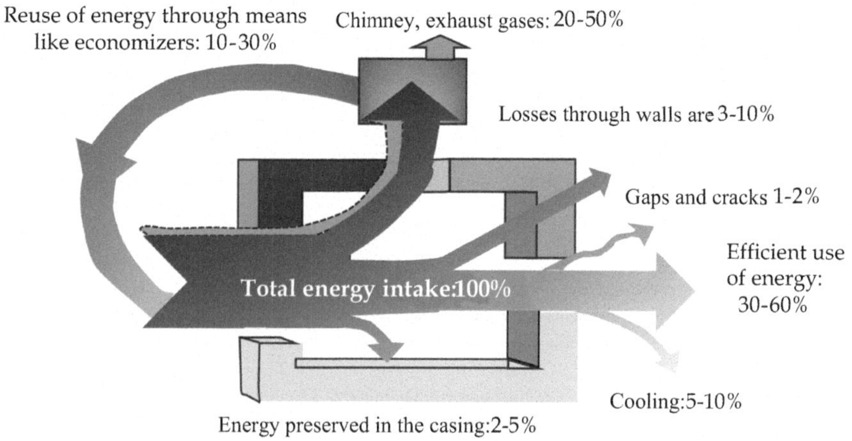

Fig. 3.2 The diagram of the EB of the company

The analysis of the EB involves qualitative and quantitative estimation of the company's energy management. It includes comparing actual energy use indicators with standards, previous period figures, future projections, and similar metrics from other companies. This comparison should consider comparable conditions (such as identical production volumes, product composition and quality, etc.).

The efficiency of energy use in an installation can be characterized by the efficiency ratio, determined by the formula:

$$h = Э_u \ / \ Э_s,$$

where $Э_u$ is the amount of energy usefully used; $Э_s$ is the amount of supplied energy.

When estimating the efficiency of energy resource use at a company, the total amount of both used and supplied energy across all installations for different types of energy carriers is considered.

As a result, studying the EB, an important efficiency indicator is estimated: the SEPC for product manufacture.

References

1. International Energy Agency (2022) World energy balances 2022. OECD. https://doi.org/10.1787/world_energy_bal-2022-en
2. Smil V (2016) Energy transitions: global and national perspectives, 2nd edn, p 296
3. Ushakov VYa (2011) Energy saving and increasing energy efficiency: socio-economic, organizational, and legal aspects: textbook. Tomsk Polytechnic University Publishing House, Tomsk, p 280
4. International Energy Agency (2019) Manual on energy balance statistics, p 227
5. Dudley B (2019) BP statistical review of world energy. British Petroleum Company, p 64

6. Bhattacharyya SC (2011) Energy economics: concepts, issues, markets and governance. Springer, p 673
7. Ürge-Vorsatz D, Herrero ST, Dubash NK, Lecocq F (2019) Measuring the co-benefits of climate change mitigation. Annu Rev Environ Resour 44:549–582. https://doi.org/10.1146/annurev-environ-101718-033102
8. Sovacool BK, Brown MA (2017) Competing dimensions of energy security: an international perspective. Annu Rev Environ Resour 42:77–108. https://doi.org/10.1146/annurev-environ-110615-085259
9. Gillingham K, Palmer K (2018) Bridging the energy efficiency gap: policy insights from economic theory and empirical evidence. Rev Environ Econ Policy 12(18):18–38. https://doi.org/10.1093/reep/rex021
10. Rakhmonov IU, Hoshimov FA, Kurbonov NN, Jalilova DA (2022) Optimization of the modes of electric loads of power-consuming units operating in different production modes. AIP Conf Proc 2552:050022-1–050022-6. https://doi.org/10.1063/5.0112391
11. Rakhmonov IU, Ushakov VYa, Najimova AM, Jalilova DA, Omonov FB (2021) Minimization of energy production management consumptions. E3S Web Conf 289:07013. In: Energy systems research. https://doi.org/10.1051/e3sconf/202128907013
12. Dincer I, Rosen MA (2015) Energy: production, conversion, storage, conservation, and coupling. Springer, p 246
13. Rakhmonov IU, Ushakov VYa, Niyozov NN, Kurbonov NN, Mamutov M (2021) Energy saving in industry. E3S Web Conf 289:07014. In: Energy systems research. https://doi.org/10.1051/e3sconf/202128907014
14. Hoshimov FA, Rakhmonov IU, Niyozov NN, Omonov FB (2022) Main regularities of change in energy indicators of energy consuming facilities of industrial enterprises. AIP Conf Proc 2552:030025-1–030025-7. https://doi.org/10.1063/5.0112388

Chapter 4
Classification of Industrial Enterprises and Technological Equipment

In this chapter, a general characterization of industrial enterprises is discussed. According to a typological classification that encompasses the main sectors of industrial production, industrial enterprises are divided into ten basic groups. The classification differentiates between industrial enterprises with continuous and discrete manufacturing cycles, as well as the equipment used in the technological process, which is categorized based on its purpose: universal, specialized, and general purpose.

4.1 General Characteristics of Industrial Enterprises

An enterprise, regardless of its form of ownership, is an important element of the country and region's economy. It acts as an independent organizational-administrative unit, whose activity is aimed at meeting public needs and generating profit.

An enterprise represents an unified production-technical complex, characterized by the common economic purpose of the produced goods, the uniformity of the technological process, and a certain qualification composition of the staff. It has the following main features:

Production-technical unity. This unity is expressed in the interconnection of technic and technology, consumed raw and material resources, forms and methods of organization of main, auxiliary, and service productions.

Organizational-administrative unity. The enterprise is a collective of workers, headed on the basis of single leadership by a manager (director), who bears personal responsibility for the results of production-economic activity, represents the interests of the enterprise at all levels of management, disposes of property and production means within the rights established by the owner, concludes contracts, approves staff

I. U. Rakhmonov et al., *Electric Consumption by Industrial Enterprises*, Power Systems, https://doi.org/10.1007/978-3-031-62676-0_4

lists, and issues orders mandatory for all employees. In turn, the workforce participates in the enterprise's activities with its labor and decides on self-management issues in accordance with the enterprise's charter.

Economic unity. The enterprise operates based on economic calculation, ensuring the creation of net income (profit), independently plans its activities, and determines development prospects, based on the available material and financial resources.

The forms and methods of organization, planning, and management of the enterprise's activities are determined by its industry affiliation, level of concentration, type of production, degree of specialization, mechanization, and automation.

Grouping enterprises into classification groups allows for each of them to develop a unified approach to organizing their activities. This applies to the formation of the production structure and management structure, principles and methods of organizing the technological process, work of auxiliary and service economies, system of operational planning and regulation of production.

An industrial enterprise—is a complex, including main production, auxiliary, energy, transport, and storage buildings. Production buildings include workshops that produce semi-finished products, parts, and finished products; auxiliary buildings include plant management buildings, household premises, canteens, etc.; energy buildings include power stations and substations, boiler rooms, compressor stations, etc. Transport and storage buildings include premises for storing raw materials, materials, semi-finished products, and finished products, and for housing transport vehicles—garages, depots, etc. [1].

Engineering structures, depending on the type of production, include storage facilities for bulk materials, tanks for liquids and gases, water supply devices in the form of water towers, cooling towers, aqueducts, and various devices for placement technological equipment.

Some productions include technological equipment in the form of open or partially open apparatus and installations, which is especially typical for metallurgical and chemical plants.

The composition of buildings and structures depends on the type of production and can be very diverse. The nature of the production process determines the necessary area for the location of the industrial enterprise and its location relative to urban development.

The diverse range of industrial enterprises can be conditionally classified into several groups based on production and technological characteristics. According to the typological classification, which covers the main branches of industrial production, industrial enterprises can be divided into the following main groups [2]:

(1) mining and mineral processing industries;
(2) metallurgical industry;
(3) petrochemical and chemical industries;
(4) machine building;
(5) instrument making and electronics;
(6) woodworking and pulp and paper industries;

(7) construction industry;
(8) textile and light industries;
(9) food industry;
(10) energy sector.

Each group includes many enterprises characterized by similar technological processes. Enterprises in each group share common features in the planning of their territories and buildings, architectural and spatial composition, and systems for engineering and social services.

In the mining and mineral processing industries, various iron and polymetallic ores, combustibles (coal, oil, gas, etc.), and non-metallic materials (salt, etc.) are extracted. These are usually located near raw material sources. These enterprises are characterized by large territories, a developed railway network, external pipelines, and numerous warehouses [3].

Enterprises in the metallurgical industry include black metallurgy plants that produce cast iron and steel, as well as non-ferrous metallurgy plants producing copper, aluminum, zinc, and other non-ferrous metals.

Black metallurgy enterprises require large areas of 600–900 ha with a workforce of 18–20 thousand people, significant energy and basic services, and a developed railway network.

In selecting locations for these enterprises, it is considered that large non-ferrous metallurgy enterprises emit a significant amount of harmful substances into the atmosphere.

The petrochemical and chemical industries include enterprises such as petrochemical, oil refining, gas processing, superphosphate, gas, nitrogen and chlorine, chemical fibers, synthetic rubber, etc. These enterprises are located near sources of raw materials. Their architectural and spatial organization is characterized by an abundance of open technological and pipeline equipment and developed railway transport. They sometimes occupy territories of 500–800 ha with a workforce of more than 10 thousand people.

In the mechanical engineering industry, heavy, medium and light engineering enterprises are distinguished by the nature of production [4].

Heavy machine-building enterprises include plants that produce metallurgical, mining equipment, excavators, etc. Medium machine-building enterprises include plants that manufacture automobiles, tractors, agricultural and road machinery, as well as machine-tool plants. Light machine-building enterprises are those engaged in the production of household appliances and electrical engineering [5].

At all mechanical engineering enterprises, the technological process includes storage of raw materials, procurement, processing and assembly of products. The location of machine-building enterprises is not tied to raw material sources. The most harmful emissions into the atmosphere are characteristic of foundry and forging-press production, and electroplating workshops. These productions require sanitary breaks of 300–500 m from residential areas. Processing and assembly productions do not emit significant amounts of harmful substances and require sanitary breaks of 50–100 m. Light machine-building enterprises are harmless and are often located within the residential territory of cities. Large plants occupy territories of more than 500 ha with a workforce of 30,000 people or more [6].

Instrument-making and electronics enterprises include plants producing vacuum electronic devices, measuring equipment, televisions, radios, tape recorders, computers, etc. These enterprises are usually located in multi-story buildings, require constant temperature and humidity control, and clean air. They occupy small territories (5–10 ha), require railway transport, and have a large workforce (from 2000 to 8000 people). They are located near the residential areas of cities [7].

Woodworking and pulp and paper industries have similar raw material sources. Woodworking includes sawmills, house-building enterprises, furniture factories, etc. Such enterprises occupy territories of 20–60 ha with up to 2000 employees, and are usually located near raw material sources. They are characterized by increased fire risk and need a developed railway transport system. Pulp and paper industry enterprises include plants for the production of pulp, paper, cardboard, etc. Large plants occupy territories of up to 300 ha with more than 6000 employees. These enterprises are characterized by the discharge of large volumes of polluted water and the emission of harmful substances into the atmosphere [8].

Enterprises in the construction industry include plants for the production of building materials—cement, gypsum, asphalt, glass, bricks, etc.; they are located near raw material sources. The construction industry also includes plants for reinforced concrete structures, house-building combines, which are located near large construction sites [9].

Enterprises in the textile and light industries produce consumer goods. Textile enterprises include cotton, wool, silk, linen, artificial fiber, etc. The territory of large combines reaches 40–60 ha, with a workforce of 8–10 thousand people. Textile enterprises are ideally located close to residential areas of cities [10].

Light industry enterprises manufacture footwear, leather goods, sewing, and knitwear products. These enterprises occupy small territories (1–4 ha), are harmless, and are situated in multi-story buildings with a large number of employees. They do not require railway connections. This creates the conditions for placing light industry enterprises directly within the residential areas of cities [11].

Enterprises in the food industry include: bakeries, confectionery factories, dairy and brewery plants, meat processing plants, etc. Sugar, oil and fat, and dairy-canning factories are built near raw material sources. Most food enterprises are located close to residential buildings, with the more harmful ones, such as cattle yards, slaughterhouses, and large meat processing plants, being located 500 m or more away from housing [12].

Energy enterprises include hydroelectric power stations (HPPs), thermal power plants (TPPs), and nuclear power plants (NPPs). HPPs are situated in zones of major rivers convenient for hydrotechnical conditions, while TPPs are built close to consumers, primarily of thermal energy—industrial enterprises and cities [13].

4.2 Classification of Industrial Enterprises with Continuous Production Cycle

Industrially enterprises are divided into two groups based on the nature of their production: (a) with a continuous and (b) discrete (intermittent) production cycle.

The discrete technological process is characterized by the separateness of main and auxiliary technological operations and the intermittency of work, set by specific norms. Enterprises in machine building, construction, mining, and other similar industries use this type of technological process. Enterprises operating in this mode lose part of their working time to auxiliary technological processes [14].

The category of industrial enterprises with a continuous production character includes metallurgical, chemical, oil refining, and textile enterprises, where main and auxiliary technological processes occur simultaneously. Enterprises with a continuous production character have several advantages: shorter production time for finished products, that is, higher labor productivity, and more complete use of main and auxiliary equipment [15].

The classification presented in Fig. 4.1 is based on materials [16–19] and is applicable only to industrial enterprises with a continuous production character.

The analysis of existing classifications shows that currently, there is no unified classifier reflecting the characteristics of enterprises operating in such a mode. The authors of this monograph have created a fairly complete classification of industrial enterprises with a continuous production nature.

In developing the classifier, the characteristics of enterprises in ferrous and non-ferrous metallurgy, chemical industry, textile enterprises, oil refining, etc., were studied in detail. It is proposed to classify such enterprises according to 12 indicators:

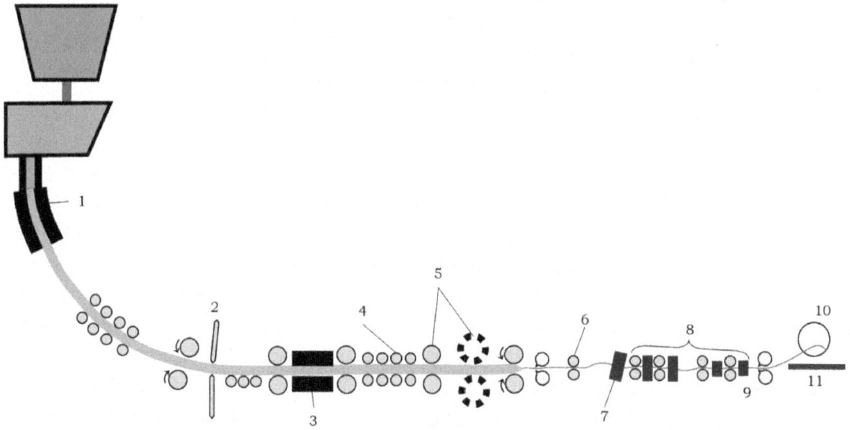

Fig. 4.1 Diagram of continuous steel sheet production: (1) continuous steel casting machine; (2) flying scissors; (3) cleaning machine; (4) induction heater; (5) planetary mill; (6) rough horizontal stand; (7) rough vertical stand; (8) block of finishing stands; (9) flying scissors; (10) coiler; (11) device for coil handling

(1) by types of sectors: industry, agriculture, transport;
(2) by volume of production: small, medium, large;
(3) by level of specialization: specialized, universal, mixed;
(4) by form of ownership and management: state, foreign, public–private partnership;
(5) by scale of product manufacturing: mass, serial, individual;
(6) by raw material consumption: processing (oil and gas processing industry, coal industry, non-ferrous metallurgy), extracting (light, food, heavy industries);
(7) by inter-industry structure: narrowly focused, multi-profile, mixed;
(8) by production process: continuous, portioned, single;
(9) by production capacity: material-intensive, capital-intensive, energy-intensive;
(10) by nature of raw material processing: analytical, synthetic, direct;
(11) by application of finished products: production materials, consumables;
(12) by stages of production: single-stage, multi-stage (preparation of individual parts of the product), complex.

4.3 Classification of Industrial Enterprises with a Discrete Production Cycle

The intermittent (discrete) production process is carried out at enterprises that produce goods whose production can be stopped at any point in the technological process without losing products or raw materials, as well as products whose manufacturing process can be timed to the duration of a shift or working day [20].

For the intermittent production process in a three-shift operation, the following schedule is recommended. In the case of intermittent production (with 41 and 36 h of work per week and five working days), the shift schedules are quite simple, as it is possible to provide a common day off for all workers at the end of the week and a break within each shift. In a two-shift operation, it is advisable for workers to be occupied in the first shift for one week and then in the second shift, followed by the first shift again [21].

In a single shift with a 41-h week, the working time is 8 h per day. Workers are occupied in the first shift for one week, then in the second, and then again in the first. To maintain an average working week of 41 h, every eighth Saturday is set as a working day [22].

If an enterprise operates in an intermittent mode, the maximum possible annual time fund is calculated based on a three-shift operation, or in the case of four shifts, a four-shift operation of the equipment, taking into account the established duration of shifts in hours. From the calculated annual fund, the standard time required for capital and planned-preventive maintenance, time for weekends and holidays, as well as non-working time in shortened shifts, pre-holiday and pre-weekend days is deducted.

If the main workshops of enterprises work in two shifts or less, then the capacity calculations are based on a two-shift work regime, and for unique and scarce equipment—based on a three-shift regime. At enterprises where the main workshops work more shifts than planned, the production capacity is calculated based on the actual working mode. In cases where equipment maintenance is carried out during working hours and this is accounted for in productivity norms, the time spent on this maintenance is not deducted from the total time fund [23].

For enterprises with an intermittent production process, the calendar time fund for the operation of equipment is taken into account based on a two-, three-, or four-shift regime. If the enterprise operates in four shifts, the shift duration is established in hours, deducting time needed for capital and planned preventive repairs, weekends and holidays, and reduced working hours on holidays [24].

For enterprises with an intermittent operating mode, the time fund for the operation of machines, aggregates, installations, and production capacity is calculated based on a three-shift (or four-shift) operating regime. If the main workshops of the enterprise work in 2 shifts (or less), then the time fund for equipment operation and production capacity values are calculated based on a two-shift working mode, and for unique and scarce equipment—a three-shift mode [25].

In workshops and sections with an intermittent production process, the operating regime of the enterprise depends on the established duration of the working day and the number of working shifts per day [26].

Some enterprises in the rubber industry with an intermittent operating mode have a reserve workforce and on technological sections of production [27].

When calculating the production capacity of enterprises and workshops with an intermittent production process, the actual annual time fund for equipment operation should be taken based on the actual number of working days per year, a 2-shift working mode with established shift durations in hours, taking into account shortened pre-holiday days, as well as time needed for repairs.

The actual annual fund of possible equipment operation time in an intermittent production process is the difference between the hours of the regime (nominal) time fund and the hours allocated for capital and planned-preventive maintenance.

To to evaluate the level of equipment utilization and calculate production capacities in sectors with an intermittent production process, the following shift coefficients of equipment operation in workshops are used: (a) normative (project) shift coefficient; (b) actual shift coefficient; (c) shift coefficient adopted in the calculations of production capacity.

Discrete production processes are carried out in a periodic mode, allowing for the operation of the enterprise in one, two, or three shifts, with production stopping on weekends and holidays. The feasibility of organizing discrete production is ensured by the fact that their interruption does not lead to negative consequences for the quality of the manufactured products, their delivery and consumption, or for safe operation.

In organizing any production process, a number of principles guided by the theory of organization are followed. These principles represent generalized, established practices and methods that have gained widespread practical application in the process of organizing any system, including the organization of production processes. The key principles of organizing a production process include the following:

Specialization. This principle of organizing production processes involves a strict division of labor within an enterprise. It entails in-plant specialization, facilitating mass production of a limited range of products in separate structural production units (workshops) of the enterprise or at workplaces of strictly defined stages of the technological process. Specialization can be object-based (on the finished product as a whole), detail-oriented (on the manufacturing of individual parts), and operation-oriented (on carrying out a specific operation of the technological process).

Production specialization ensures, on one hand, an increase in its efficiency, and on the other hand, can lead to negative consequences. An increased level of specialization results in improved economic indicators due to the increased output of homogeneous products, including based on the emergence of broader automation opportunities in production processes, increased labor productivity of workers performing the same strictly specialized functions, and due to the improved quality of manufactured products. However, specialization is often associated with the monotony and uniformity of workers' functions, leading to increased psychological and physical stress, potential de-skilling, loss of interest in work, and consequently, a decrease in labor productivity and staff turnover.

The production program of the enterprise determines the level of in-plant specialization; it is influenced by factors such as standardization, rationing, and unification of product designs, and typification of technological processes and their parameters. The principle of specialization and its adherence largely determines the successful implementation of other principles of organizing production processes.

Discrete manufacturing—it is a type of production in which the raw material is transformed into the final product through a finite number of technological and assembly operations [28].

Typically, the start and end of these operations are determined by signals from two-position sensors. Discrete manufacturing is predominant in industries such as machine building, instrument engineering, light industry, and in enterprises producing furniture, packaging, pharmaceuticals, and others.

According to the International Association of Engineering Technologists, this type of production exists in more than 75% of industrial enterprises worldwide. Even in industries where production is exclusively continuous, discrete processes are present as auxiliary components.

The principles of such work organization can be illustrated using the example of a commonly found type of enterprise with discrete manufacturing. In such an enterprise, an automated system ensures control, sorting, accounting, and formation of shipping batches, final inspection, and the creation of a database for component products arriving at the enterprise using Radio Frequency Identification (RFID) tags. RFID– is a method of automatic object identification, where data stored in so-called transponders (*transmitter-responder*), or RFID tags, are read or written via radio signals.

Discrete manufacturing is characterized by the intermittency of the production process throughout its entire length. This intermittency complicates the movement of the material flow from the system's input to its output, which is reflected in the increased duration of the production cycle, an increase in the volume of all types of inventory, and ultimately leads to a decrease in the system's productivity. Such production requires constant monitoring of all processes to ensure the required productivity.

4.4 Classification of Technological Equipment

Enterprises of the considered type face not only the task of ensuring the continuity of the technological process but also the challenge of increasing the efficiency of EPC. The relevance of the second task is driven by the high energy intensity of these enterprises (significant electricity consumption), which was noted earlier. In this context, the method of EPC management should provide solutions to both of these tasks.

Achieving this can be done by changing the operating mode of the electro-technological equipment. In this regard, two main methods can be used: (1) switching off the electro-technological equipment at times and for durations dependent on its place in the technological process; (2) changing the operating modes of the equipment without disrupting the main technological process.

In the selection and implementation of the method, information about the role and place of specific electrical equipment in the technological process and its composition is important.

Considering this, the authors of the monograph have refined the classification of technological equipment used in the technological process of this type of industrial enterprises. It is proposed to divide this equipment into two main groups, and each of these into subgroups, as shown in Fig. 4.2 [29, 30].

Two main groups include, respectively, main and auxiliary equipment.

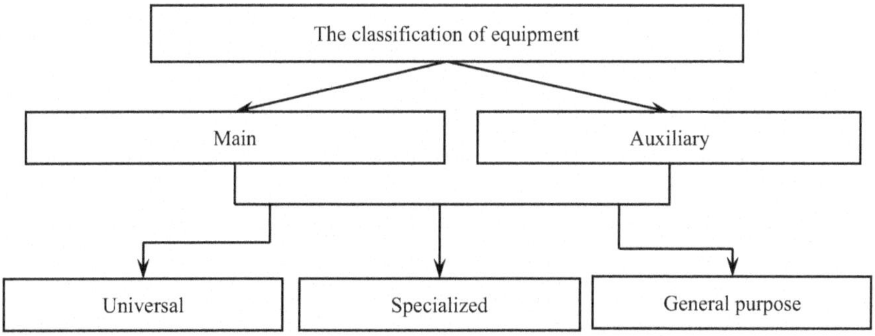

Fig. 4.2 Classification of equipment in enterprises with continuous production nature

The main group equipment is used at all stages of the technological process—from raw material acquisition to the production of finished products. This equipment constitutes 70–80% of all industrial equipment of the enterprise.

The share of ***auxiliary equipment*** in the electrical load of an industrial enterprise is about 20%. In managing the consumption of electricity by auxiliary technological equipment, the focus should be on managing the energy consumption of group equipment. It is known that due to the special place of auxiliary equipment in the technological process, disconnecting it from the power grid for a certain period does not disrupt the continuity of the technological process. Considering the low power consumption of each element of this equipment, it is advisable to use group regulation in managing EPC.

All equipment used in the technological process is classified by purpose into three groups [6]:

(1) universal;
(2) specialized;
(3) for general purposes.

Universal equipment is used at all stages of the technological process and is highly efficient. It can be used not only in the main but also in the auxiliary technological processes.

Specialized equipment is used only in certain types of special processes, the content of which depends on the nature of the technological process at a given enterprise.

General purpose equipment is used to perform auxiliary operations of technological processes of various productions.

Managing electrical power consumption with auxiliary equipment may not provide sufficient efficiency. To save electrical energy, it may be necessary to disconnect some of the main electricity power consumption of technological equipment from the power supply system. However, in this case, the following requirements must be met:

- The use of equipment in the technological process should not be necessary,
- Disconnecting the equipment should not cause significant damage to the enterprise,
- Disconnecting this equipment should not lead to the disconnection of other equipment in the same group, i.e., it should not disrupt the continuity of the technological process.

The third requirement is the most stringent, as its non-compliance can lead to the disconnection of a whole range of other equipment connected to it, and, as a consequence, to significant economic damage to the enterprise.

The classification of equipment—it is a scientifically justified distribution of machines and devices into separate groups based on certain characteristics. For instance, for chemical production equipment, the most important criteria are production purpose, the process carried out in the machines or devices, functional purpose, the principle of organization of the technological process, the degree of automation, etc. Depending on the production purpose, it is divided into the following classes:

- technological,
- energy,
- transport, repair,
- lifting,
- auxiliary.

Technological equipment, based on the nature of the processes occurring in it, is divided into the following classes:

- for mechanical processes;
- for hydro—mechanical processes;
- for thermal processes;
- for mass transfer processes;
- for chemical processes.

Within each class, equipment is subdivided into groups according to functional purpose. Moreover, each group of equipment is divided into types, and types into standard sizes. Auxiliary equipment is intended for implementing auxiliary production processes, in particular, the storage and transportation of liquids, liquefied gases, vapors, and gases, bulk materials in volumes and directions required by the technological process. Auxiliary equipment for chemical production includes pumps and compressors, tanks for storing liquids and liquefied gases, gas holders, bunkers, and silos for bulk materials, etc.

Technological equipment, based on the nature of its impact on raw materials, semi-finished products, or products, can be divided into machines and devices.

A machine is a mechanical device consisting of parts working in coordination, performing specific purposeful movements to transform energy, material, or information.

In accordance with this definition, three classes of machines are distinguished:

- engine machines;
- technological machines;
- computing machines.

Machines and devices, based on the principle of organization of the technological process, are divided into continuous-action equipment and periodic-action equipment.

References

1. Despeisse M, Ball PD, Evans S, Levers A (2016) The industrial ecology of the automobile manufacturing sector. J Clean Prod 116:299–305. https://doi.org/10.1016/j.jclepro.2015.12.080

2. International Energy Agency (2017) Energy efficiency. OECD Publishing, p 143. https://iea.blob.core.windows.net/assets/2eed56ff-591a-412a-82f8-5f7eb61831a1/EEMR2017_web.pdf
3. Laurence D (2015) Establishing a sustainable mining operation: an overview. J Clean Prod 2(24):275–284. https://doi.org/10.1016/j.jclepro.2015.02.024
4. Rajput RK (2015) A textbook of manufacturing technology: manufacturing processes. Laxmi Publications, p 854
5. Introduction to manufacturing processes. McGraw-Hill Education. ISBN: 978-0070311367
6. Groover MP (2018) Automation, production systems, and computer-integrated manufacturing. Prentice Hall, p 811
7. Seitz JL, Wells PE (2016) Global issues: an introduction. Wiley. ISBN: 978-1118968857
8. Schmidt R, Möhring M, Härting RC, Reichstein C, Neumaier P, Jozinović P (2015) Industry 4.0—potentials for creating smart products: empirical research results. Bus Inf Syst Eng 67(3):261–278. https://doi.org/10.1007/s12599-015-0383-3
9. Gibson GE, Gebken RJ (2015) Introduction to construction materials: Managing the supply chain for high-performance building materials. J Constr Eng Manag 141(1):04014052. https://doi.org/10.1061/(ASCE)CO.1943-7862.0000929
10. Pal R, Gander J (2018) Competitiveness of the EU textile and clothing industry in the global value chain. Res Int Bus Financ 42:469–476. https://doi.org/10.1016/j.ribaf.2017.07.089
11. Gam HJ, Cao H, Farr C, Heine L (2016) The business of fashion: designing, manufacturing, and marketing, 5th edn. Fairchild Books, p 234
12. Gunden N, Moros E, Shane A (2018) The impact of local food environments on consumers' food purchasing behaviors. Int J Food Sci Article 4717504
13. Sovacool BK (2016) The national and international energy security risks of fossil fuels and nuclear power. In DellaSala DA, Goldstein MI (eds) The encyclopedia of the Anthropocene, vol 3, pp 169–178
14. https://mydocx.ru/12-61522.html
15. https://econ.wikireading.ru/30202
16. https://studfile.net/preview/5565925/page:2/
17. http://stroy-spravka.ru/article/osnovnye-opredeleniya-tipologicheskaya-klassifikatsiya-promyshlennykh-predpriyatii
18. https://libraryno.ru/1-4-2-klassifikaciya-promyshlennyh-predpriyatiy-ekon_org_proizv/
19. https://studopedia.org/10-33084.html
20. Allaev KR (2021) Modern energy and prospects for its development. Fan va texnologiya, p 923 (in Russian)
21. Allaev KR (2020) Modern scenarios for the development of the world energy industry. Problems Energy Sources Saving 3–4:14–24 (in Russian)
22. Allaev KR (2017) The role of "green" energy in the energy balance of the world and Uzbekistan. Problems Energy Sources Saving 3–4:11–29 (in Russian)
23. Allaev KR (2006) Energy of the world and Uzbekistan. Moliya, p 388 (in Russian)
24. European Commission (2019) Directive 2019/944 on common rules for the internal market in electricity and amending directive 2012/27/EU. Official J Eur Union 158:125–196 (Available at: EU Legislation Portal)
25. International Energy Agency (IEA) (2020) Policy pathway: improving the efficiency of energy use in the industrial sector. IEA Publications, July 2020. Available at: [IEA Website]
26. United Nations Industrial Development Organization (UNIDO) (2016) Guidelines for accelerating the replacement of outdated and inefficient industrial equipment. UNIDO Reports, December 2016. Available at: [UNIDO Publications]
27. Johnson D, Thompson S (2011) Integrated solutions for energy conservation in the metallurgy industry. J Ind Energy Eng 4(1):35–42
28. Williams R et al (2014) Energy-efficient electric power management in steel manufacturing. Chief Energy Eng J 5:10–20

29. Ushakov VY, Mytnikov AV, Rakhmonov IU (2023) High-voltage equipment of power systems. In: Power systems. Springer, Cham. https://doi.org/10.1007/978-3-031-38252-9
30. Zhang H, Li H (2016) A review on the integration of supply chain management and industrial big data analytics for improved manufacturing. Int J Prod Res 54(5):1297–1311

Chapter 5
Changes in Energy Parameters Over Time

This chapter is focused on analyzes typical daily load profiles of electrical power consumption (EPC) in industrial enterprises of various profiles, based on multi-year data. The peculiarities of the technological processes at these enterprises are described, and the main patterns of (EPC) by these businesses are determined. In the process of analyzing the dynamics of electrical loads, the main problems in managing (EPC) under conditions of variable demand are identified, and priority directions for solving this issue are determined.

5.1 Dynamics of Electrical Loads

The analysis of electrical loads and the dynamics of (EPC) by industrial enterprises begins with the study of their electrical load graphs. All industrial enterprises have certain features in their daily and annual electric load profiles, conditioned by differences in technological characteristics.

Below, the graphs of electrical load of enterprises in dynamics are investigated. The analysis showed that the current standard electrical load graphs for enterprises in the oil refining, chemical, ferrous and non-ferrous metallurgy, and textile industries do not take into account technologies and modes introduced in recent years. Therefore, for a number of industrial enterprises of this type, electrical load graphs (Figs. 5.1, 5.2, 5.3, 5.4 and 5.5), have been developed and presented in the monograph, taking into account innovative developments and peculiarities of technological processes.

I. U. Rakhmonov et al., *Electric Consumption by Industrial Enterprises*, Power Systems, https://doi.org/10.1007/978-3-031-62676-0_5

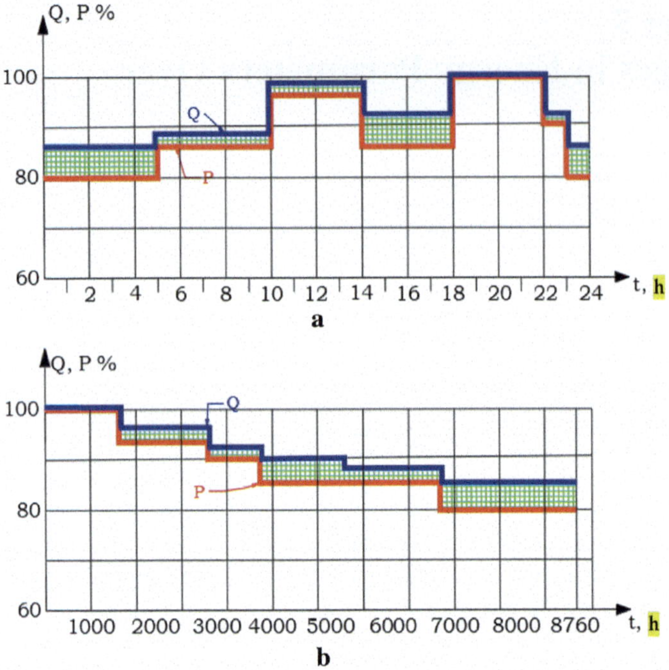

Fig. 5.1 Daily (**a**) and annual (**b**) graphs of active and reactive electrical loads in oil refining industry enterprises

5.1.1 Oil Refining Industry Enterprises

Oil refining plants (ORPs) are energy-intensive enterprises that belong to one of the most important sectors of the economy, contributing to solving tasks related to energy security and economic transformations at the state level. It should be noted that, due to the peculiarities of the technological process at these enterprises, there is no possibility to use traditional technologies for managing electrical loads [1, 2].

From the data in Fig. 5.1, it follows that it is not possible to suspend the oil refining process even for a short period. However, reductions in load at peak power can be achieved by modifying the auxiliary technological process modes.

The electrical load graphs play a crucial role in selecting rational operating modes for the electrical equipment in refineries, improving the efficiency of EPC, forecasting electrical power demand, and designing optimal power supply for the enterprise. The graphs demonstrate the stability of the electrical load at the refinery. The daily graph in Fig. 5.1a shows that the enterprise managed to minimize electrical load consumption during the morning peak, but in the evening peak, the energy consumption reached its maximum.

In the manufacturing of one ton of product, the specific electrical power consumption (SEPC) averages 60–80 kWh/ton, and in the most energy-intensive technological devices, this value is 100–150 kWh/ton. In the refining margins (crack spread)

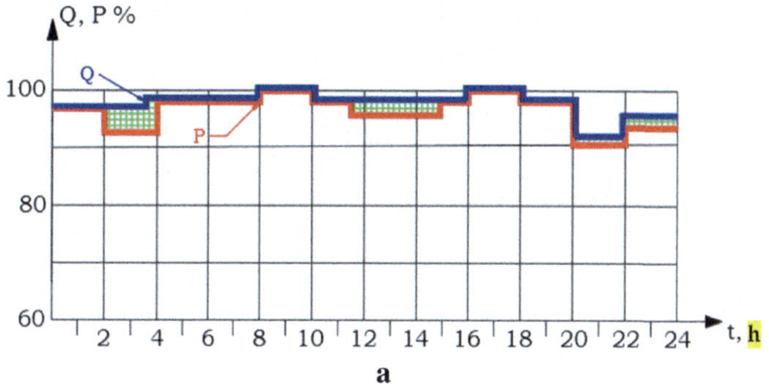

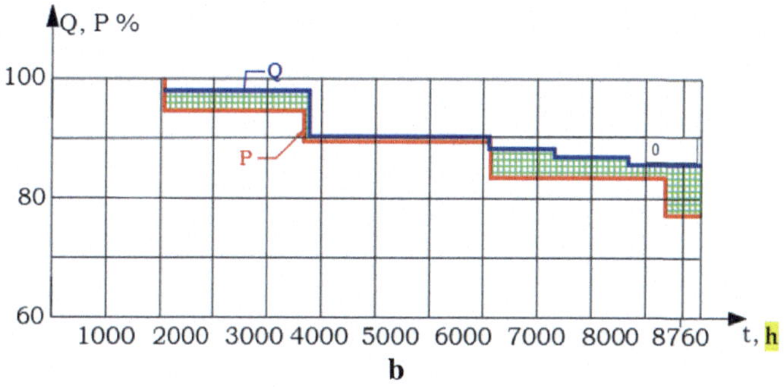

Fig. 5.2 Daily (**a**) and annual (**b**) graphs of active and reactive electrical loads in the chemical industry

process at the refinery, this indicator is lower by 5–7%. However, under a forced technological regime, this value increases by 7–10 times, significantly exceeding the reduction in the cost of 1 kWh of consumed electrical power during the load minimum in the electrical grid. The economic impracticality of such a managerial decision is evident [3, 4].

The implementation of new technology in oil refining has enabled the maintenance of the production level of light petroleum products while reducing the volume of processed raw materials.

5.1.2 Chemical Industry Enterprises

The chemical industry—is one of the most energy-intensive sectors of the economy in developed countries. The rapid development of the chemical industry leads to an increase in fuel, heat, and EPC, outpacing the overall industrial energy consumption growth due to the high-energy intensity of chemical production processes. A characteristic feature of the chemical industry is the wide variety of products manufactured, the presence of different technological processes, and energy consumers,

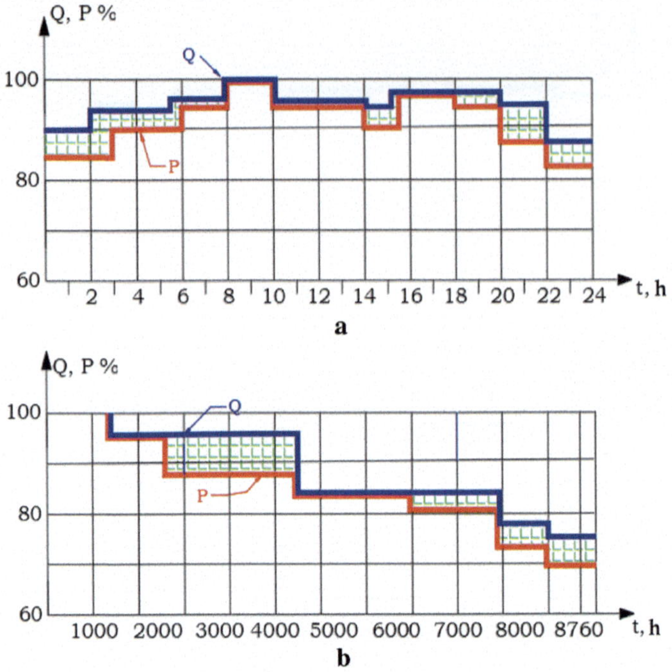

Fig. 5.3 Daily (**a**) and annual (**b**) graphs of active and reactive electrical loads in ferrous metallurgy enterprises

which poses unique requirements for energy utilization in each case. Chemical enterprises are notably characterized by high consumption of thermal energy in the form of technological steam and hot water, used both in production processes and for heating and sanitary needs. For instance, the share of EPC is 47%, while the share of thermal energy reaches 53% of the total energy consumption [3, 5].

Continuous chemical productions require uninterrupted supply of electrical and thermal energy. Interruptions in energy supply can cause accidents, accompanied by significant material damage and time expenditures for the restoration of the interrupted technological process. In addition to underproduction of goods and failure to meet production plans, this leads to unproductive energy resource expenditures for reheating, warm-ups, starts, and more, which increase the losses of enterprises.

The mentioned characteristics of chemical productions necessitate a flexible approach to the management of EPC, taking into account the specifics of the technological process of producing each type of product.

In terms of energy intensity, the chemical industry also surpasses many other economic sectors. If in the cost of finished products manufactured by industrial enterprises, the cost of electrical energy is on average 2.5%, in the chemical industry, this indicator is 9%. The most energy-intensive productions include the manufacturing of ammonia, phosphorus, calcium carbide, sodium carbonate, and other similar products [3, 6].

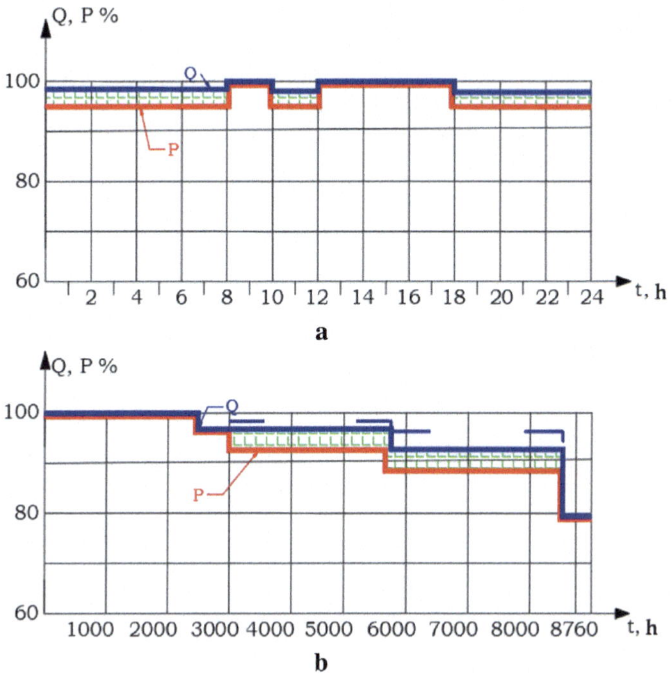

Fig. 5.4 Daily (**a**) and annual (**b**) charts of active and reactive electrical loads of non-ferrous metallurgy enterprises

In many chemical industry enterprises, technological processes using large magnitude direct current are applied. Currently, the main sources of such current are motor-generators (rotating converters) and, in rare cases, outdated mercury rectifiers. Lately, new power semiconductor rectifiers are increasingly being used. Their advantages are ease of operation and higher efficiency. Rectifier units are built for several tens of thousands of amperes, therefore their efficiency gains lead to a noticeable reduction in EPC.

An analysis of the balance of EPC at chemical enterprises shows that the main part of the electrical power consumed by the enterprise is used in electrochemical, electrothermal, and electromagnetic processes. These processes, in turn, consist of technological processes such as solution preparation, solution electrolysis, heating, melting, high-temperature synthesis, ore processing, and other similar processes. Along with the main technological equipment, a significant part of the electrical power is consumed by auxiliary equipment, primarily electric motors of various purposes.

Energy-intensive chemical enterprises require uninterrupted supply of electrical energy and heat. In sectors specializing in the production of scarce and expensive goods, disruption of the technological process due to interruptions in the power

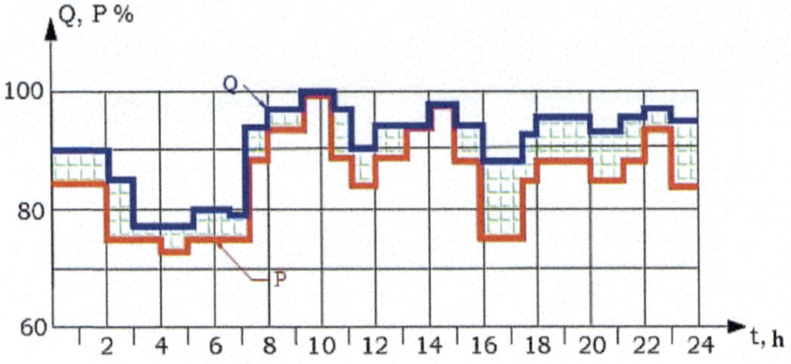

a

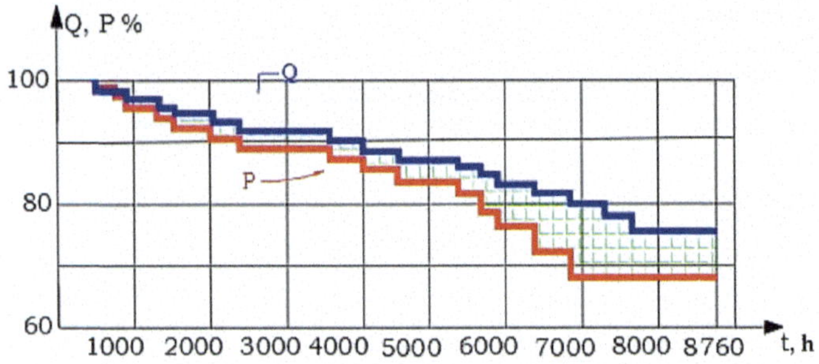

b

Fig. 5.5 Daily (**a**) and annual (**b**) charts of active and reactive electrical loads of textile industry enterprises

supply is accompanied by significant negative consequences not only for the specific enterprise but also for the economy on a national scale. An analysis of the daily and annual load graphs of a chemical plant (Fig. 5.2) shows that its power consumption is relatively stable—load fluctuations on average do not exceed 5–7%.

5.1.3 Enterprises in Ferrous Metallurgy

In current metallurgy, it is difficult to delineate the boundaries between production and energy processes: technological and energy transformations, based on scientific achievements in thermal power engineering, electrification, chemistry, physics, and electronics, are closely intertwined. The energy sector of ferrous metallurgy encompasses areas such as thermification and electrification, provision of compressed air, oxygen, nitrogen, gasification of production, fuel usage, water supply, purification of water and exhaust gases, and systems for communication and automation of production processes.

Electric arc furnaces in ferrous metallurgy enterprises consume about 1 billion kWh of electrical energy annually on average, making them one of the largest consumers of electrical energy. The energy departments of ferrous metallurgy enterprises ensure not only the main and auxiliary technological processes (water supply, water and exhaust gas filtration, compressed air production, communication system operation, etc.), but also the supply of raw materials (nitrogen, argon, oxygen, water) and the implementation of environmental protection measures (water filtration, exhaust gas purification, etc.) [7–9].

When operating a ferrous metallurgy plant in three shifts, the daily electrical load graph takes the form shown in Fig. 5.3a. As noted above, the steelmaking shop is the largest consumer of electrical power at the enterprise, with an average daily electrical power consumption of 3–4 million kWh. The maximum annual electrical load in winter is 10–15% higher than in summer, which can be explained not only by climatic factors but also by the periodicity of demand for products, as shown in Fig. 5.3b.

It is also observed that the minimum value of the required active power in terms of duration reaches 70%, which is associated with the untimely delivery of primary raw materials and the performance of a large volume of intermediate repairs at the enterprise [10–12].

5.1.4 Non-ferrous Metallurgy Enterprises

Non-ferrous metallurgy, with its unique energy consumption characteristics, is a leading industrial sector in many developed countries having the appropriate raw materials. The specificity of its EPC is defined by the diversity of its cycles, the inseparable connection between technological processes, and the need for large volumes of secondary energy resources. The industry, on average, consumes about 15% of the total electrical energy produced, with total energy costs in the cost of some products accounting for 35 55%.

The optimal structure of energy supply in the industry can only be determined based on a technical–economic analysis and the development of measures to reduce energy costs in production. It is necessary to reconsider the approach to the issue of supplying the country's economy with energy resources, i.e., to consider this issue comprehensively—not only from the perspective of increasing fuel extraction and energy production but also in terms of saving them on a national scale. The significant energy intensity and high level of power supply in production in non-ferrous metallurgy stimulate the fight for the rational use of traditional energy resources. Given the energy-technological nature of non-ferrous metal production, issues of energy resource conservation are among the most important for collectives of power engineers and technologists.

Research results show that energy-intensive consumers in non-ferrous metallurgy enterprises include workshops for the production of aluminum, copper, nickel, lead–zinc alloys, and titanium-magnesium. In this context, the production of 1 ton of aluminum (including the production of alumina) consumes about 9 tons of conventional fuel, nickel—13.4 tons, zinc—2 tons, copper—1.4 tons, and lead—0.9 tons [3, 13, 14].

In the industrial sector, companies experiencing significant overconsumption of energy resources are visited by teams of technologists and energy specialists. These teams are tasked with identifying the reasons for this overconsumption and developing measures to eliminate it. The energy saving efforts in the industry are focused on the following key areas:

- reducing direct energy resource consumption at all stages, maximizing the use and integration of secondary energy resources into the active energy balance;
- transitioning to new technological processes that ensure comprehensive utilization of raw materials based on closed-loop, waste-free material and energy balances;
- minimizing losses in the conversion of all forms of energy.

The daily load graph of the non-ferrous metallurgy enterprises, Fig. 5.4a demonstrates minor fluctuations in electrical load. This stability is attributed to the continuity of production processes at the enterprise and the timely delivery of primary raw materials. From the annual load graph, Fig. 5.4b, it is evident that the reactive power consumed by the enterprise exceeds its active power, indicating the presence of specific electrical equipment at the facility.

5.1.5 Textile Industry Enterprises

Leading manufacturers in this sector include cotton, silk, and knitwear-sewing enterprises. The production of cotton and silk fabrics is particularly energy-intensive.

Apart from the main technological process equipment, significant EPC in textile enterprises is also characteristic of equipment for ventilation, heating, and lighting. The primary energy resource expenditures in the textile industry include the operation of equipment, finishing processes, ventilation, heating, and lighting.

Each type of fabric production encompasses three main stages—spinning, weaving, and finishing, each differing significantly in energy intensity. In the textile industry, spinning is the most energy-intensive, while finishing is the most heat-intensive. Consequently, the greatest potential for electrical power savings (EPS) is in the spinning-weaving processes, and for thermal energy, in the finishing process.

At first glance, the low power of textile enterprise equipment does not categorize them as energy-intensive. However, the multistage nature of the production process leads to high energy resource consumption. The production of each product consists of three main phases, varying greatly in specific energy intensity: spinning, weaving, and finishing.

The production process at textile enterprises is organized in 2–3 shifts; the schedule of electrical loads is variable. For spinning, weaving, and dyeing-finishing processes producing cotton fabrics, there is a relative consistency of electrical loads during working shifts. However, for many textile industry enterprises, typical electrical load graphs do not reflect the specific operation modes of the equipment.

An analysis of the electrical load graphs, as shown in Fig. 5.5, reveals that the EPC of spinning and textile factories is specific and variable. Comparing the daily electrical load graphs of the textile industry with those of other sectors helps identify the unique characteristics of this industry. The variation in the electrical load graphs of textile enterprises averages around 20%. Significant fluctuations in this indicator can also occur due to changes in the technological process.

A distinctive feature of the textile industry enterprises is the inability to predict their EPC with acceptable accuracy. This is primarily due to the diversity of types and quality of raw cotton, which is the main raw material in the textile industry.

Thus, the study of electrical loads and EPC by textile enterprises has identified the following characteristics of EPC dynamics:

- the charts show minor dips in electrical load, associated with the absence of raw materials and fluctuations in production volume;
- there is a decrease in EPC during the summer period and an increase in the autumn period, reflecting the characteristics of continuous technological production;
- on all enterprises, the electrical loads on daily charts fluctuate within $\pm$ 20%, indicating stability and uniformity of finished product manufacturing;
- uniform distribution of electrical loads ensures minimal fluctuations in the specific electrical power consumption for production (within $\pm$ 5–6%);
- there are abrupt changes in EPC, which are uncharacteristic for such enterprises;
- the instability of electrical loads disrupts the regularities of EPC and does not allow for full implementation of modeling, standardization, and forecasting procedures, taking into account the influence of main and auxiliary factors.

5.2 Management Challenges in the Context of Variable Electrical Power Consumption

The challenges of enhancing energy efficiency in industrial enterprises are primarily due to the lack of a unified methodology and approaches for their resolution. The development of such methodologies is complicated by the characteristics of equipment operating in continuous modes, which are marked by complex and multilevel technological processes. The tasks that need to be addressed to overcome this problem can be conditionally grouped into the following categories:

- studying the patterns of change and management of EPC modes depending on the nature of the technological process;
- determining the enterprise's prospects in terms of EPC based on normative and forecast indicators, calculated through advanced standardization and forecasting of EPC;
- development and analysis of energy balance;

- optimization of equipment operating modes throughout the technological process based on criteria of minimal EPC and maximal production efficiency.

Solving the tasks grouped in these categories is key to addressing the main challenge—enhancing the efficiency of EPC for all types of production processes.

There are several internal and external factors whose direct influence or consideration can improve energy efficiency. These include production, technological, meteorological, and other factors. Mathematical modeling must account for these factors and the patterns of their changes.

In the modern economy, mathematical models and algorithms based on single-factor analysis cannot provide sufficient accuracy. Moreover, the change in these factors and their impact on the considered indicator are not described by a single-factor mathematical model. The development of mathematical models based on multi-factor analysis that combines all elements is necessary.

The primary methods for increasing energy efficiency in industrial enterprises, which are already in use and need further refinement and wider application, include the following:

- improving the system for accounting and controlling EPC in industrial enterprises;
- constructing energy characteristics of technological equipment and selecting optimal operating modes;
- developing scientifically substantiated mathematical models for standardization and forecasting of EPC with high accuracy based on multi-factor analysis and considering the nature of the technological process.

It is known [15–18] that the analysis of the EPC of any industrial enterprise, its efficiency, and the determination of its targeted use is carried out by studying the patterns of its changes. It is during these studies that the opportunity arises to determine the level and dynamics of EPC. As mentioned earlier, the technological process of industrial enterprises has its own peculiarities due to its complexity. Therefore, forming the laws of EPC is also a complex task. When making certain decisions, these patterns are revealed through the analysis and study of EPC for different periods. These patterns are subsequently used not only in the analysis, regulation, and forecasting of EPC but also serve as the primary source of information in the design of power supply for such enterprises.

When studying the patterns of EPC of industrial enterprises, the volume and duration of the initial data and the stability of the state of this data for each stage of the enterprise's operation are of great importance.

The operating modes of equipment in industrial enterprises vary depending on production and technological conditions; therefore, for most types of electrical equipment, it is not possible to directly measure energy indicators. In such cases, statistical processing of primary data is conducted, focusing on the following:

- in-depth study and analysis of the nature of EPC by the equipment under study, and the formation of laws for a series of time intervals;
- determination of parameters influencing electrical power consumption;
- assessment of the accuracy of measuring instruments;

- determination of primary data over a period of not less than three years to obtain reliable results in the processing of statistical data;
- exclusion of rapidly changing data and data acting over short time intervals from the primary data set.

Currently, the process of standardizing EPC is based on the method of determining normative indicators at each level of consumption, depending on the location of consumers. The overall factory normative indicators are determined based on the final generalization of these indicators.

In refining the methodology for standardizing EPC at industrial enterprises, it is important to solve the following tasks:

- determining scientifically substantiated normative indicators of EPC, considering the interrelationship between EPC and the main factors of the production process;
- studying the energy parameters of technological equipment and corresponding measurements, on the basis of which normative energy characteristics of energy-intensive equipment will be constructed;
- determining normative indicators of specific EPC, taking into account the seasonality of the enterprises' operation (autumn–winter and spring–summer);
- considering production factors that affect specific EPC, such as lighting, temperature-humidity regime;
- including the composition and quality of the primary raw materials in the calculations.

In the process of developing models for forecasting EPC, it is necessary to consider not only the absolute values of the indicators but also the patterns of change in technological and energy indicators. Improvements in the technological process affect not only the efficiency of equipment operation but also the SEPC.

Analysis of the issue of improving the efficiency of EPC by industrial enterprises shows that for workshops, departments, production, and technological processes, a unified methodological approach, and theoretical foundations for studying the laws of EPC are necessary.

References

1. Allaev KR (2020) Modern scenarios for the development of the world energy industry. Probl Energy Sources Sav 3–4:14–25 (in Russian)
2. Allaev KR (2017) The role of "green" energy in the energy balance of the world and Uzbekistan. Probl Energy Sources Sav 3–4:11–29 (in Russian)
3. Smith JE (2018) Advanced energy characterization for industrial electrical power consumption analysis. In: Industrial energy management: trends and applications, vol 1, no 3. Springer, pp 85–89
4. Lisa N, Brown M (2019) Innovative strategies for energy savings in the industrial sector. In: Energy efficiency series. Elsevier, p 320

5. Maria G, Wilson R (2020) Standardization of fuel and energy resource expenditures in modern industrial enterprises. J Ind Sustain 13(2):22–24
6. Chen X et al (2019) Multilevel standardization and forecasting methods for industrial electrical power consumption. J Electr Eng 15(4):47–56
7. Harris J et al (2018) Comprehensive energy conservation solutions in contemporary metallurgical industries. Metall Ind Press 3:50–56
8. Patel AK (2019) Strategies for the rapid modernization of aging industrial equipment. Ind Dev Perspect 2:34–40
9. Claire J, Grant E (2020) New approaches to electrical power consumption standardization and forecasting in industry. Ind Energy J 4(1):18–23
10. Samuel L, Cho K (2020) Advancements in neural network-based electrical power consumption forecasting for industrial applications. Energy AI J 2(3):10–15
11. Wei Z, Yang L (2021) Artificial neural network applications in energy consumption forecasting. In: Energy forecasting studies, vol 2, no 4. CRC Press, pp 102–108
12. Kim D et al (2018) Neural network techniques for short-term energy consumption forecasting in industrial sectors. J Appl Energy 21(5):72–80
13. Gnatyuk VI et al (2020) Optimal control of power consumption of the regional electrical complex using rank analysis methods: educational and methodological manual. Publishing House of the IKBFU. I. Kanta, Kaliningrad, p 233 (in Russian)
14. Rakhmonov IU, Nematov LA, Niyozov NN, Reymov KM, Yuldoshev TM (2020) Power consumption management from the positions of the general system theory. J Phys Conf Ser 022054. https://doi.org/10.1088/1742-6596/1515/2/022054
15. Rakhmonov IU, Reymov KM (2019) Regularities of change of energy indicators of the basic technological equipment of the cotton-cleaning industry. J Phys Conf Ser 1399:055038. https://doi.org/10.1088/1742-6596/1399/5/055038
16. Rakhmonov IU, Reymov KM (2019) Mathematical models and algorithms of optimal load management of electricity consumers. Energetika Proc CIS High Educ Inst Power Eng Assoc 62(6):528–535. https://doi.org/10.21122/1029-7448-2019-62-6-528-535
17. Hoshimov FA, Rakhmonov IU, Niyozov NN, Omonov FB (2022) Main regularities of change in energy indicators of energy consuming facilities of industrial enterprises. AIP Conf Proc 2552:030025-1–030025-7. https://doi.org/10.1063/5.0112388
18. Rakhmonov IU, Hoshimov FA, Kurbonov NN, Jalilova DA (2022) Optimization of the modes of electric loads of power-consuming units operating in different production modes. AIP Conf Proc 2552:050022-1–050022-6. https://doi.org/10.1063/5.0112391

Chapter 6
Analysis of Energy Indicators

A detailed analysis of the energy indicators of industrial enterprises is necessary due to the complexity of technological processes, which does not allow for a comprehensive analysis of these indicators. In light of this, the monograph provides a detailed description of the methodology for analyzing energy indicators based on decomposition, as proposed by the authors of the monograph. This complex task is resolved in a simpler manner by dividing the analyzed parameters into several subsystems. Additionally, a method for estimating the share of electrical power consumption (EPC) for auxiliary needs in the calculations of the energy intensity of each type of product is proposed; this ensures high accuracy in determining the desired indicators.

6.1 Energy Indicators Analysis Based on Decomposition

Identifying patterns in the changes of the energy intensity of produced goods and determining the conditions under which EPC indicators may significantly change over time are essential for ensuring accurate forecasting. Additionally, the reliability of forecasting is influenced by the development of new and the modernization of existing technologies and equipment, which vary across different industrial sectors. The energy intensity of products and the level of EPC also depend on changes in the structure of electricity expenditure; this, in turn, is influenced by the demand for components of the technological process such as compressed air, oxygen, nitrogen, etc.

Consequently, a set of tasks related to accounting for these factors can be formulated, the resolution of which is necessary for improving the accuracy of forecasting.

The projected total electrical power consumption of an enterprise includes the following components:

I. U. Rakhmonov et al., *Electric Consumption by Industrial Enterprises*, Power Systems, https://doi.org/10.1007/978-3-031-62676-0_6

- EPC by the enterprise when, in the considered forecast period, the equipment is not replaced, and traditional technologies are used (W_b);
- EPC by the enterprise after modernization of technological equipment and transition to new technology (W_m);
- EPC by a new enterprise, with new equipment and technologies (W_n).

$$W = [V_b(\beta_b d_b) + V_m(\beta_m d_m) + V_n(\beta_n d_n)], \quad (6.1)$$

where V_b, V_m, V_n—volumes of planned products;

β_b, β_m, β_n—proportions of the planned types of products manufactured by the enterprise;

d_b, d_m, d_n—correspondingly, the SEPC specific electrical power consumption for each type.

In accordance with the prospective development plan of an enterprise, such as in the black metallurgy sector, the planned production volumes for the forecasting period (V) and the proportions of existing, modernized, and new productions (β_b, β_m, β_n). are determined. When determining prospective specific electrical power consumption (d), factors such as the type and technical condition of the main technological equipment in the melting process, metal nomenclature, the volume of consumed components in the technological process, and secondary energy carriers (compressed air, oxygen, etc.) are considered. Errors in improperly accounting for these factors can lead to significant deviations from real values of electrical power consumption. The monograph proposes a method, illustrated by an enterprise in the metallurgical sector, that enables the determination of expected specific electrical power consumption with sufficient accuracy. Energy indicators are examined in the process of scaling up and accelerating work on the rational use of energy resources. This process particularly emphasizes determining the dependence of product energy intensity on electrical, technological, and operational parameters, considering their specific characteristics [1]. A systemic analysis is utilized, taking into account the diversity of interconnected factors and events at various scales: unit, workshop, and enterprise.

The features of the object are identified, taking into account the operating mode and production characteristics of the investigated object and factors influencing energy intensity indicators. In particular, [2]:

- operating mode over time;
- constant mode of continuous operation;
- variable mode;
- characteristics of production (uniform, homogeneous, heterogeneous);
- components (material resources) consumed in the technological process (compressed air, argon, oxygen, etc.);
- presence of inter-industrial connections;
- implementation of measures to save energy resources.

Based on the analysis, mathematical models of specific EPC are developed for each type of product (by divisions and the entire enterprise), taking into account the

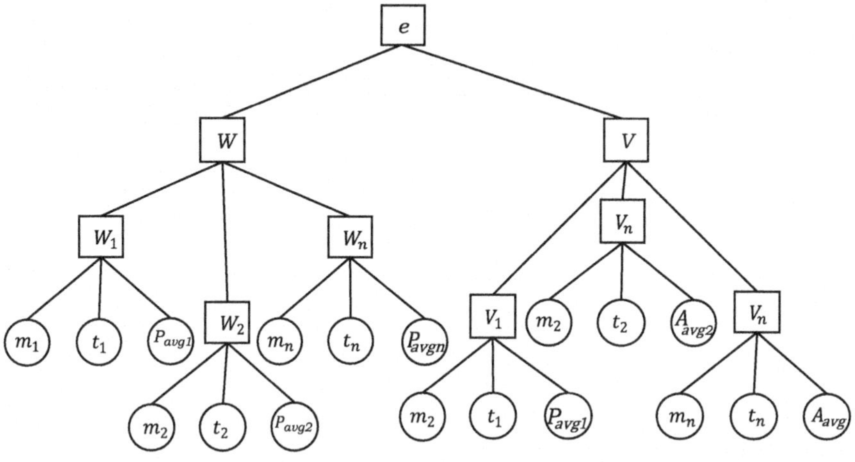

Fig. 6.1 Scheme of decomposition of SEPC indicators

influence of relevant factors. Considering the actual volumes of EPC by workshops, which may vary over time due to changes in the volumes of produced products, its electrical capacity is determined.

To identify reserves for saving electricity from the overall calculation scheme, the power of units participating in the manufacturing of products during the maximum load on the power system is highlighted. Schemes of decomposition are developed for objective assessment of EPC on separately for the workshop, enterprise, and association, according to the approved scenario.

At the beginning, the decomposition of specific electrical power consumption (SEPC) indicators is performed. Then, the main subtasks are identified, the solution of which ensures the achievement of the set goal, as shown in Fig. 6.1.

Overall, in the scheme of decomposing the indicators of SEPC (e) of the unit producing diverse products ($P_1, P_2, \ldots, P_n$) with hourly productivity ($A_1, A_2, \ldots, A_n$), depicted is the consumed power of the unit (V), average power (P_{avg}) for each type of product ($P_{\text{avg}1}, P_{\text{avg}2}, \ldots, P_{\text{avg}n}$), power of equipment for auxiliary needs (P_{aux}), equipment operating time ($t_1, t_2, \ldots, t_n$), number of operating units (m) for each type of product ($m_1, m_2, \ldots, m_n$), as well as the EPC (W) for the production of each type of product ($W_1, W_2, \ldots, W_n$).

By applying decomposition schemes, SEPC indicators can be determined using the following formulas. The EPC (W_i) for the production of goods is equal to:

$$W_i = m_i \cdot t_n \cdot P_{\text{avg}i}, \tag{6.2}$$

where m_i—the number of operating units for the production of each type of product;

$P_{\text{avg}i}$—average power for each type of product, kW;

t_n—equipment operating time, hours.

The volume of production for each type of product (V_i) is determined by the expression (6.3):

$$V_i = m_i \cdot t_n \cdot A_{avgi}, \tag{6.3}$$

where: A_{avgi}—hourly productivity, *t/h*.

As an example, let's determine the indicators of SEPC for the Electric Arc Furnace Shop (EAFS) (see Table 6.1) using the decomposition scheme, Fig. 6.1.

We determine the daily EPC (W_i) for the production of each type of product using formula (6.2):

Angle 40 × 4—$W_1 = m_1 \cdot t_1 \cdot P_{avg_1} = 6 \cdot 16 \cdot 56 = 5376$ kWh;
Angle 32 × 4—$W_2 = 6 \cdot 16 \cdot 49 = 4704$ kWh;
Square 16 × 16—$W_3 = 4 \cdot 16 \cdot 43 = 2752$ kWh; (eliminate the variety in writing lines here and on page 123).
Square 20 × 20—$W_4 = 4 \cdot 16 \cdot 52 = 3328$ kWh;
Flat bar 40 × 5—$W_5 = 8 \cdot 16 \cdot 31 = 3968$ kWh;
Flat bar 50 × 5—$W_6 = 8 \cdot 16 \cdot 36 = 4608$ kWh;

We determine the volume of production for each type of product (V_i) using the formula (6.3):

Angle 40 × 4—$V_1 = m_1 \cdot t_1 \cdot A_{cp_1} = 6 \cdot 16 \cdot 7.2 = 691.2$ t;
Angle 32 × 4—$V_2 = 6 \cdot 16 \cdot 6.6 = 633.6$ t;
Square 16 × 16—$V_3 = 4 \cdot 16 \cdot 6.4 = 409.6$ t;
Square 20 × 20—$V_4 = 4 \cdot 16 \cdot 6.8 = 435.2$ t;
Flat bar 40 × 5—$V_5 = 8 \cdot 16 \cdot 4.6 = 588.8$ t;
Flat bar 50 × 5—$V_6 = 8 \cdot 16 \cdot 5.1 = 652.8$ t;

Table 6.1 Main production indicators

№	Product range name	Number of machines, units	Operating time, hours	Average productivity, A_{cp} tons/hour,	Average power consumption P_{cp}, kW
1	Angle 40 × 4	6	16	7.2	56
2	Angle 32 × 4	6	16	6.6	49
3	Square 16 × 16	4	16	6.4	43
4	Square 20 × 20	4	16	6.8	52
5	Flat bar 40 × 5	8	16	4.6	31
6	Flat bar 50 × 5	8	16	5.1	36

We determine the specific electrical power consumption for each type of product (e_i):

Angle 40 × 4—$e_1 = \frac{W_1}{V_1} = \frac{5376}{691.2} = 7.8\,\text{kWh/t}$;
Angle 32 × 4—$e_2 = \frac{4704}{633.6} = 7.4\,\text{kW/t}$;
Square 16 × 16—$e_3 = \frac{2752}{409.6} = 6.7\,\text{kW/t}$;
Square 20x20 $- e_4 = \frac{3328}{435.2} = 7.64\,\text{кВт}\cdot\text{ч}/\text{т}$;
Flat bar 40 × 5—$e_5 = \frac{3968}{588.8} = 6.74\,\text{kW/t}$;
Flat bar 50 × 5—$e_6 = \frac{4608}{652.8} = 7\,\text{kW/t}$.

Based on the above, we find the EPC for the calculation period [2].

$$W_c = K\left[W_{\text{aux}}(1 + f_i) + \delta \cdot A_{\text{avg}}(1 + l_i)\right], \tag{6.4}$$

where

K—number of cycles in the calculation period;
f_i, l_i—respectively, the resulting values of factors influencing the EPC for auxiliary needs and productivity;
W_{aux}—EPC for auxiliary needs of the unit per cycle;
A_{avg}—average hourly productivity of the unit per hour;
δ—a coefficient representing the constant component of SEPC.

In addition to the unit-level scenario, scenarios are also developed for higher levels of management—enterprises, associations, and the dispatch center of the power system. These scenarios, according to the existing scheme (task tree), should be developed for each type of production, both similar and different kind of products [3, 4]. If an enterprise, in addition to the main products, manufactures goods for supply to other enterprises, then the EPC for the manufacture of these types of products should be standardized separately. Moreover, the calculation should include the EPC for the production of auxiliary components of the technological process (ACTP), manufactured by auxiliary workshops [5, 6].

6.2 Estimation of the Proportion of Electrical Power Consumption for Auxiliary Needs in the Calculations of Electrical Intensity for Each Type of Product

In the electric balance (EB) of an enterprise, the EPC for auxiliary needs and losses independent of the efficiency of the technological process are highlighted. To obtain the total electrical intensity for each type of product, the proportions of energy expenditure on these needs should be added to the workshop-specific indicators of SEPC [7–9]. The complexity of analysis and calculation is in the fact that the components of auxiliary needs (compressor and pumping stations, etc.) produce their products for each workshop and, accordingly, for each product. A method for distributing these resources among types of products is proposed [10, 11]. This task can be solved based on a block-differentiated method, which allows aggregating the SEPC for the production of generated resources (d) and the specific consumption of these resources per unit of semi-finished or finished product (q). Thus, the proportion (β) can be determined by the formula:

$$\beta = d \cdot q, \tag{6.5}$$

where: q—the normative amount of the corresponding raw materials or resources consumed per unit of semi-finished or finished product.

The SEPC for each type of product, respectively, will be:

$$e = d_c + \beta, \tag{6.6}$$

where: d_c—SEPC for the production of the i-th type of conditional product.

Currently, in determining the participation of ACTP, their stages in the production of the final product are not considered. The same component (compressed air B_c, argon B_a, oxygen B_k etc.) is used at several stages of product manufacturing (see Fig. 6.2). For instance, the EPC for the production of compressed air is:

$$W_c = \omega_c \left(B_{c_1} + B_{c_2} + \cdots + B_{c_n} \right). \tag{6.7}$$

The calculation formula for the block of the production unit, considering the energy indicators of auxiliary components, is as follows:

$$d_i = \varepsilon_1 \cdot q_1 + \sum_1^m \varepsilon_2 \cdot q_2, \tag{6.8}$$

where m—the number of ACTP blocks used in the production of goods by the workshop; $q_2 = \frac{B_i}{V_i}$—the specific consumption of this type of ACTP per unit of product V_1.

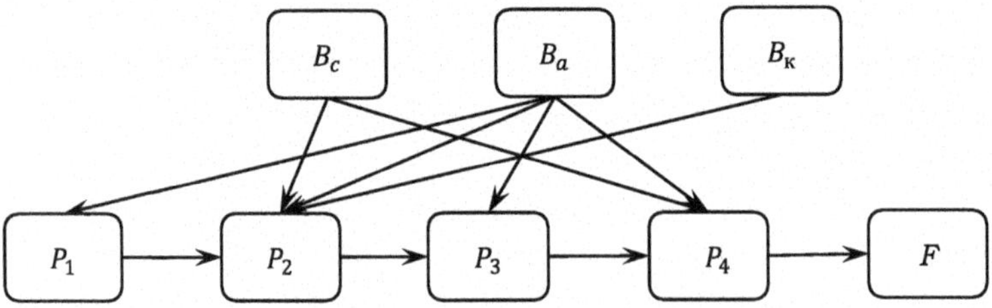

Fig. 6.2 Scheme for provisioning workshops providing with ACTP

As an example, the general production SEPC considering the energy indicators of the ACTP, is determined according to Fig. 6.2. At the Electric Arc Furnace Shop (EAFS), compressed air, oxygen, nitrogen, and argon are used as auxiliary components in the technological process. Below is an example of the calculation for compressed air. Oxygen compressor 82,000 m^3/h, electric arc furnace—100 – $B_m{}^C = 48{,}542$ km^3/year, steel treatment complex unit—$B_o{}^C = 28{,}542$ thousand m^3/year. Total power consumption of the compressor station—$P_n = 9450$ kW; volume of final product output—$F = 626{,}700$ m^3/year.

We determine the SEPC for the production of compressed air in the electric arc furnace shop (EAFS):

$$\omega_c = \frac{9450}{48,542} = 194\,\text{kWh/1000}\,\text{m}^3.$$

The consumption of compressed air per unit of final product will be:

for DSP should be written full term before—100 – $q^c_{DSP-100} = \frac{48{,}542{,}000}{626{,}700} = 0.08\,\text{m}^3\text{/t}$;

for AKOS should be written full term before—$q^c_{AKOS} = \frac{28{,}542{,}000}{626{,}700} = 0.045\,\text{m}^3\text{/t}$.

Specific energy consumption per unit of final product in production will be:

for DSP – 100 – $\alpha^c_{DSP-100} = 194 \cdot 0.08 = 15.52\,\text{kWh/t}$;

for AKOS – $\alpha^c_{AKOS} = 194 \cdot 0.045 = 8.73\,\text{kWh/t}$.

Total energy consumption for producing a unit of the final product will be:

$$e_{EPCC} = \alpha^c_{DSP-100} + \alpha^c_{AKOS} = 15.52 + 8.73 = 24.25\,\text{kWh/t}.$$

We determine the total energy consumption for compressed air production using formula (6.5):

$$W_o = 194 \cdot (485{,}42 + 28{,}542) = 14954\ \text{тыс.кВт} \cdot \text{ч} / \text{год}.$$

The proposed method allows for calculations at each level of the production structure. The factors described by the given equations, which determine the forecasting object, enable the analysis of the initial indicator and reveal the interconnections between energy and technological factors, see Fig. 6.3. Each of these parameters, in turn, is associated with a range of factors that directly or indirectly influence them. Based on experimental research and statistical analysis of operational materials, we identify the factors that have the most significant impact on energy indicators [12–14].

The set task is solved by constructing the relationship of standard energy characteristics W with the volume of produced goods V. Based on these energy characteristics, the multifactorial correlation model is described as follows:

$$e = f(X_1, X_2, \ldots, X_n), \tag{6.9}$$

where: $X_1, X_2, \ldots, X_n$—factors in correlation with the energy intensity of products (Table 6.2).

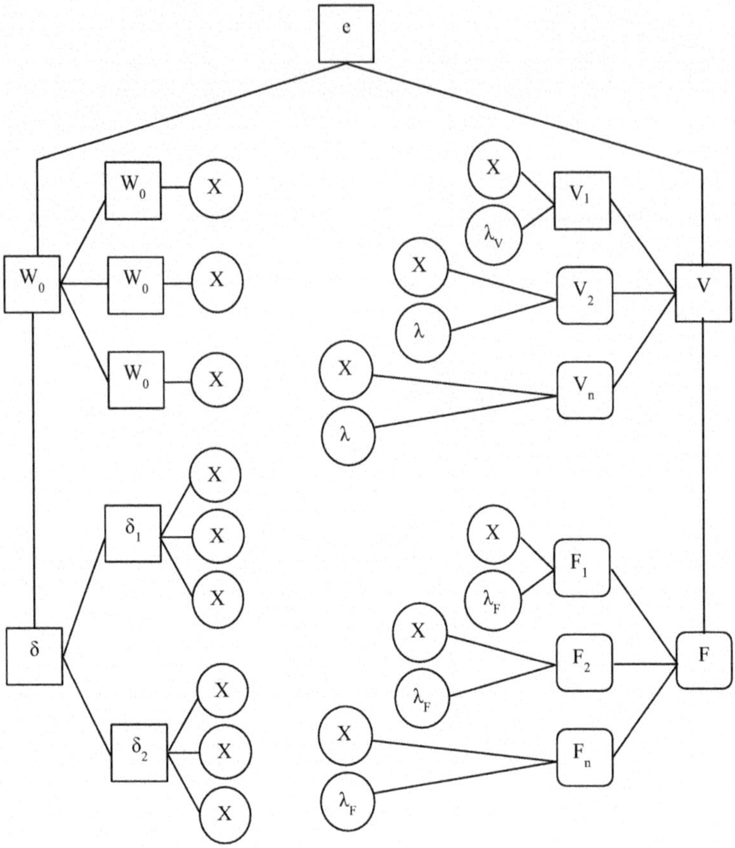

Fig. 6.3 Task tree for determining specific electrical power consumption, divided into a series of subsystems (blocks)

Similar models are developed for the entire set of indicators.

The constant component of EPC (independent of productivity) is equal to:

$$W_o = W_{o_1}(1 \pm k_1) + W_{o_2}(1 \pm k_2) + W_{o_3}(1 \pm k_3)\ ; \\ k = f(X_0, X_1, X_2), \tag{6.10}$$

where

W_{01} is the EPC for lighting and ventilation;
W_{02}—is the EPC for air conditioning;
W_{03}—is the EPC for the idle running of machines and mechanisms;
X—represents influencing factors: X_0—seasonality, X_1—ambient temperature; X_2—periodicity of machines and mechanisms maintenance.

Table 6.2 Factors in correlation with the energy intensity of products

Indicators	Characteristic structural components of the indicator	Influencing factors
Constant component of EPC (independent of productivity), W_o	Electric lighting W_{01}, ventilation, air conditioning W_{02}, idle running of machines and mechanisms W_{o3}	Seasonality X_0, ambient temperature X_1, frequency of maintenance of machines and mechanisms X_2
Partial SEPC, δ	Specifically utilized electrical power δ, specific load losses of electrical power	Temperature of processed material X_3, humidity X_4, granularity X_5, hardness X_6, pressure X_7, equipment load factor influencing workshop production volume X_8
The volume of workshop production: natural, V; equivalent, V';	Product range: First product V_1, second product V_2, n-th product V_n	Electrical intensity of individual types of products: $X_{9_1}, \lambda_{a_1}, X_{10_2}, \lambda_{a2}, X_{10_n}, \lambda_{a_n}$
Volume of factory production (Final): F	Product range: First item F_1 Second item F_2 n-th item F_n	Electrical intensity of individual items, influencing the volume of produced product types: $X_{11_1}, \lambda_{z_1}, X_{11_2}, \lambda_{z_2}, X_{11_n}, \lambda_{z_n}$
Processed product (raw material), G	Processed product (raw material), G	Content of useful product X_{12}, extraction rate of useful product X_{13}
ACTP, B	Compressed air B_c, oxygen B_κ, argon B_a	Network pressure X_{14}, losses in units X_{15}, network losses X_{16}, temperature X_{17} and others X_n

The partial SPEC δ:

$$\delta = \frac{W - W_o}{V}(1 \pm k_{\text{aux}});\ \delta_{\text{aux}} = \frac{W_{01} - W_{02}}{V}(1 \pm k_{\text{aux}});$$
$$k_{\text{aux}} = f(X_4, X_5, X_6, X_7, X_8, X_9). \tag{6.11}$$

Volume of workshop natural production V:

$$V = (V_1 + V_2 + \cdots + V_n)(1 \pm k_q)\ ;$$
$$k_q = f(X_{10_1}, X_{10_2}, \ldots, X_{10_n}), \tag{6.12}$$

where: V represents volume of products of different assortments: the first product V_1, the second product V_2, the n-th product V_n;

$X_{9_1}, \lambda_{a_1}, X_{10_2}, \lambda_{a_2}, X_{10_n}, \lambda_{a_n}$—electrical intensity of individual types of products;

Volume of factory production (Final) F:

$$F = (F_1 + F_2 + \cdots + F_n)(1 \pm k_z)\ ;$$
$$k_z = f(X_{11_1}, X_{11_2}, \ldots, X_{11_n}), \tag{6.13}$$

where F represents the product range: F_1—the first item; F_2—the second item, F_n—the *n*-th item.

$X_{11_1}, \lambda_{z_1}, X_{11_2}, \lambda_{z_2}, X_{11_n}, \lambda_{z_n}$—electrical intensity of individual items, influencing the volume of manufactured products of various types.

Auxiliary components of the technological process, $\textit{в}$:

$$\begin{aligned} \textit{в} &= \textit{в}_1 + \textit{в}_2 + \cdots + \textit{в}_n\,; \\ \textit{в} &= f\left(X_{14}, X_{15}, X_{16}, X_{17}, X_n\right), \end{aligned} \tag{6.14}$$

where: $\textit{в}_1$—compressed air, $\textit{в}_2$—oxygen, $\textit{в}_3$—argon;

X_{14}—network pressure, X_{15}—losses in units; X_{16}—network losses, X_{17}—temperature; X_n—and others.

In forecasting calculations, it is necessary to consider the planned introduction of new machines and technologies, automation tools, labor protection measures, etc. The energy evaluation of these factors can be carried out using known methods, including the expert evaluation method [9].

In conclusion, it should be noted that a systematic approach in analyzing energy indicators allows for a comprehensive examination of the trends and patterns in the change of specific and absolute EPC, taking into account the combined influence of electrical, technological, and operational factors, thereby ensuring the reliability of calculations.

References

1. Brown John E, Smith David A, Nguyen Lisa K (2019) Exploring the energy characteristics of periodically operated equipment: case study of advanced DSP technologies. In: Proceedings of the international scientific and practical conference on youth achievements in science for industry. Energy Section, London, pp 60–63
2. Smith J, Zhang L (2020) Comprehensive assessment of energy metrics in manufacturing industries. J Ind Energy Manage 36(2):157–174
3. Johnson E, Patel S (2021) Evaluating energy efficiency parameters in industrial sectors. Energy Ind Rev 45(1):89–103
4. Khoshimov FA (2006) Optimizing the use of energy resources in the textile industry. In: Fan, p 252 (in Russian)
5. Thompson R, Garcia M (2019) Energy performance analysis in heavy industries: methodologies and findings. Int J Energy Res 54(4):236–250
6. Nguyen H, Lee K (2022) Energy indicators and their role in sustainable industrial development. Sustain Ind Pract 12(3):200–215
7. Brown A, Green D (2018) A comparative study of energy consumption patterns in European industrial enterprises. Eur J Energy Econ 29(6):415–429
8. Rakhmonov IU, Niyozov NN (2019) Optimization setting of steel-smelting industry in the issue of alloy steels. Web conference, vol 139. Rudenko International Conference "Methodological problems in reliability study of large energy systems" (RSES 2019) 139:01077. https://doi.org/10.1051/e3sconf/201913901077
9. Rakhmonov IU, Reymov KM, Najimova AM, Uzakov BT, Seytmuratov BT (2019) Analysis and calculation of optimum parameters of electric arc furnace. J Phys Conf Ser. APITECH-2019 1399:055048. https://doi.org/10.1088/1742-6596/1399/5/055048

10. Dzeventskiy AYa, Khoshimov FA, Ibragimov KH (2000) Multivariate solution of problems of analysis, forecasting and rationing of electricity at industrial enterprises producing diverse products. Ind Energy (6):43–46 (in Russian)
11. Hoshimov FA, Rakhmonov IU, Niyozov NN, Omonov FB (2022) Main regularities of change in energy indicators of energy consuming facilities of industrial enterprises. AIP Conf Proc 2552(030025):1–7. https://doi.org/10.1063/5.0112388
12. Rakhmonov IU, Taslimov AD, Khoshimov FA (2016) Calculation of energy consumption standards by type of output of metallurgical production shops. Electronic Sci J "Research in Technical Sciences" Russia, Moscow 1(19):3–6 (in Russian)
13. Rakhmonov IU, Hoshimov FA (2020) Development of an algorithm for evaluating the dominant factors that have the greatest impact on the energy intensity of products. E3S web conference, vol 209. Sustainable Development and Smart Management, ENERGY-21, 07018. https://doi.org/10.1051/e3sconf/202020907018
14. Rakhmonov IU, Reymov KM (2019) Regularities of change of energy indicators of the basic technological equipment of the cotton-cleaning industry. J Phys Conf Ser APITECH-2019 1399:055038. https://doi.org/10.1088/1742-6596/1399/5/055038

Chapter 7
Rationing of Electrical Power Consumption

The chapter deals with identifying and analyzing factors that affect electrical power consumption (EPC). It offers methodologies for determining normative values of EPC based on multifactor analysis and on the criterion of minimizing production costs. The proposed methods are substantiated based on real calculations for operating enterprises.

7.1 General Provisions

Research into energy consumption by industrial enterprises has revealed the presence of untapped the potential of energy saving and improving energy efficiency (ES and IEE). Its realization largely depends on the proper organization and technical–economic justification of rationing EPC. Scientifically substantiated standards for specific EPC (SEPC) provide a foundation for calculating the electricity needs of production areas, workshops, and the entire enterprise.

Industrial enterprises operate based on progressive rationing, the aim of which is to determine the SEPC for specific production conditions, ensure rational and economical consumption of electricity in the production process, and establish the real electricity needs of production. The SEPC norm refers to the maximum permissible amount of electricity necessary for producing a unit of established quality product, taking into account planned organizational and technical production conditions.

The methodically correct definition of the SEPC norm forms the foundation for developing in-plant standards, segmented by production processes and differentiated in the directions (articles) of expenditure and centers of losses [1].

The SEPC serves as an indicator characterizing the energy intensity of the entire technological process, a stage of production, an individual production operation, or the entire production. It is largely determined by the volume of production of enterprises and their energy—consuming equipment, the SEPC is simultaneously a

I. U. Rakhmonov et al., *Electric Consumption by Industrial Enterprises*, Power Systems, https://doi.org/10.1007/978-3-031-62676-0_7

derivative of the technological process, as it is organically connected with its modes, rhythm, raw material characteristics, and other technological factors. On the other hand, the stability of the technological process is determined both quantitatively and qualitatively by the influencing factor.

Thus, the presence of direct and inverse relationships in technology and energy in industrial enterprises is a factor determining not only technological and economical but also purely technological indicators.

As previously emphasized, the study and rationing of SEPC in the technological process deserve special attention. Often, there is a need to develop in-production norms, differentiated by items of EPC and loss centers. The main principle of compiling such norms is their segmentation by production processes. This allows for the control of EPC at all production stages, especially those with increased EPC, and identifies production areas where electrical power is efficiently used. Such a standard determines the costs of producing a unit of product, based on the permissible level of technology and organization of the technological process at the enterprise. It is composed of costs for technology, expenditures for auxiliary needs, as well as losses of electrical power in networks and transformers [2].

With changes in equipment operating modes, the SEPC norms essentially become the sole indicator for objectively estimating EPC. Finally, SEPC norms are vitally important for planning EPC for the current period and the future. Using special energy indicators, it is not difficult to determine the need for electrical power if the volume of planned production is known.

The SEPC norm—is a complex synthetic indicator, linked with many factors related to energy supply, technology, and organization of production. It depends on the technical condition of the equipment, technological indicators of production processes, equipment operating modes, the production site schedule, and many other aspects of the organization of production and operation of the enterprise's energy management [3–5].

The energy services of enterprises still permit inaccuracies in the rationing, estimation, and distribution of energy resources, as they do not fully take into account the specifics of the production process, the EPC modes of technological equipment, and the peculiarities of its operation. In this context, it is necessary to improve the methods of analysis and prediction of EPC by manufacturing enterprises and their associations, which represent self-accounting objects with independent balance and reporting.

7.2 Identifying Factors Influencing Electrical Power Consumption

In determining scientifically justified norms for SEPC to ensure high accuracy in identifying an enterprise's electrical power needs, it is essential to consider both primary and auxiliary influencing factors. Depending on the specifics of the technological process, methods of probability theory and mathematical statistics are employed to determine the degree of their impact on SEPC. The SEPC is represented as a polynomial of the form:

$$Y = a_0 + a_1x_1 + a_2x_2 + a_3x_3 + \ldots + a_nx_n, \quad (7.1)$$

where,

$Y = d$—SEPC for producing final products, kWh/ton,
$a_0, a_1, a_2, a_3, \ldots, a_n$—coefficients of the regression equation,
$x_0, x_1, x_2, x_3, \ldots, x_n$—influencing factors.

According to the analysis of existing approaches to this task, all influencing factors (energy, technological, operational, etc.) are conventionally divided into two groups: primary and auxiliary.

The primary factors affecting SEPC include characteristics of the technological scheme, equipment's technological regime, productivity, product quality, raw material quality, equipment's technical condition, and its load degree [6–8].

The auxiliary factors encompass aspects such as seasonality, random fluctuations in product and raw material quality, random deviations from technology, changes in equipment condition, and environmental parameter changes [9, 10].

In calculating absolute and SEPC, their dependencies on hourly productivity and the volume of produced goods are used. However, other factors, both random and episodic, can significantly influence the consumed power and energy. These include ambient or processing product temperature, moisture and granularity of processing products, material hardness, etc. [11, 12].

Mathematical statistical methods, with goodness-of-fit criteria, allow for an objective evaluation of energy consumption under conditions of changing product assortment and volume, and assess the impact on encompass aspects EPC indicators of energy, technological, and operational factors. Along with examining the influence of each factor individually, it is essential to evaluate their cumulative impact. Established methods for determining the characteristics of objects, based on inputting test signals or accounting for changes in the magnitude of the aforementioned production factors, have significant limitations, as often-operational conditions do not allow for the alteration of any factor due to the risk of disrupting technological conditions [13, 14].

Research has established that among all existing methods for solving this task, the stepwise regression method is most suitable. It is computationally efficient and incorporates the advantageous aspects of the method of all possible regressions and the method of elimination. It also allows for the generation of an adequate mathematical model of the object under study, suitable for practical application [15, 16].

The implementation of the exclusion method begins with the most comprehensive equation, encompassing all variables, and involves the sequential reduction of the number of variables until a decision is made to use the equation with the remaining terms. The method allows achieving the same results by working in the reverse direction, that is, by sequentially including variables into the equation until it becomes satisfactory. The order of inclusion is determined using the partial correlation coefficient, as a measure of the importance of variables not yet included in the equation [17].

This method enables the determination of multiple variable regression based on linear regression dependencies, and the restructuring of the correlation matrix ensures high accuracy of calculations, as the matrix is constructed step-by-step to the final value. The calculation of the desired value using this method starts with the construction of the correlation matrix. We select a variable X_i (for instance, x_i) that is most highly correlated with Y_i and find the first-order linear regression equation $\hat{Y} = f(x_i)$. Then we determine the partial correlation coefficient $x_i (i \neq 1)$ (taking into account the correction for x_i). Mathematically, this is equivalent to determining the correlation between the residuals from the regression $\hat{Y} = f(x_i)$ and the residuals from another regression—$X = f(x_i)$. Next, a variable x_k (assume, x_2) is chosen, which possesses the aforementioned properties, and as a result, a second regression equation is obtained $\hat{Y} = f(x_i, x_k)$.

This process continues. At each stage, the variables included in the equation at previous stages are additionally investigated. A variable introduced into the equation at an earlier stage may later turn out to be superfluous due to the relationship between this and other variables contained in the equation.

For verification, at each stage, it is necessary to calculate the partial F_1 criteria for each variable of the equation and compare them with a pre-selected significance equation F_T (F value is taken from the table). If $F_i < F_T$, then the i-th variable is excluded from the equation, and the regression equation is recalculated considering the remaining variables. Otherwise, the equation remains unchanged. This relationship allows for assessing the contribution that the i-th variable can make to the equation. A variable contributing insignificantly is excluded from the equation. This process is continued until all insignificant variables are excluded from the equation.

The construction of the mathematical model includes the following stages:

- A table with dimensions $\{n, m\}$ is compiled from the values of factors influencing the studied energy indicator (e, W), where n—is the number of measurements by factors, and m—is the number of factors determined by the method of expert evaluations.
- The existing values of each factor are summed up:

$$S_m = \sum_{i=1}^{n} X_i, \tag{7.2}$$

$$S_m = \sum_{i=1}^{n} Y_i, \tag{7.3}$$

where:

m—is the number of factors,
n—is the number of measurements of influencing factors,
Y_i—is the studied energy parameter (e, P, W).

The average value is determined by the following expression:

$$\overline{X}_r = S_m/n, \overline{Y}_r = S_{m+1}/n. \tag{7.4}$$

The standard deviation is equal to:

$$\sigma = \frac{\sqrt{\sum_{i=1}^{n} (X_i - \overline{X})}}{n}. \tag{7.5}$$

The corrected sum of squares is calculated as:

$$P_x = \sum_{i=1}^{n} X_1^2 - \left(\sum_{i=1}^{n} X_i\right)/n, \tag{7.6}$$

$$P_y = \sum_{i=1}^{n} Y_1^2 - \left(\sum_{i=1}^{n} Y_i\right)/n. \tag{7.7}$$

The uncorrected sum of mixed products is equal to:

$$S_i = \sum_{i=1}^{n} X_i Y_i. \tag{7.8}$$

Correlation coefficients are calculated between each factor and the response:

$$r_{x_1, x_2} = \frac{\sum X_1 X_2}{\sqrt{(\sum X_2^1)(\sum X_2^2)}}, \tag{7.9}$$

$$r_{xy} = \frac{\sum (X_1 - \overline{X})(Y_1 - \overline{Y})}{\sqrt{\sum (X_1 - \overline{X})^2 \sum (Y_i - \overline{Y})^2}}. \tag{7.10}$$

A matrix of correlation coefficients is constructed. The correlation matrix has the dimension $m \times m$, that is:

$$R = \begin{bmatrix} r_{11}, r_{12}, \ldots, r_{1m} \\ r_{21}, r_{22}, \ldots, r_{2m} \\ \ldots \\ r_{m1}, r_{m2}, \ldots, r_{mm} \end{bmatrix}. \tag{7.11}$$

The correlation matrix R—is expanded as follows:

$$A = \begin{bmatrix} R(m \times m) & T'(m \times 1) & I(m \times m) \\ T(1 \times m) & S(1 \times 1) & O(1 \times m) \\ -I(m \times m) & O(m \times 1) & O(m \times m) \end{bmatrix}, \tag{7.12}$$

where:

$R\ (m > m)$ is the correlation matrix for m—independent variables,
$T(1 > m)$—is the correlation vector for m—independent variables with the deviation Y,
$T(1 > m)$— the transformed vector T,
$S(1 \times 1)$—the correlation coefficient of the response with itself, $I(m \times m)$—the identity matrix,
$I(m \times m)$—the negative identity matrix. We denote the elements of this matrix as A_{ij}.

To introduce an influencing factor into the regression model, we calculate V_i using the formula:

$$V_i = r_{iy} \cdot r_{iy1}/r = A_{n-1} \cdot A_{k-1}/A_n, \tag{7.13}$$

where: $i = 1 \div k$.

We select V_i with the maximum value; let's assume this to be V_L, then X_L is the factor to be considered.

Using the table, we determine the standard F_T—criterion with a confidence probability $P = 0.95(n - 1)$. If X_L is the first variable to be included in the regression equation, its value according to the analysis of variance table is determined as follows:

- for the overall correlation coefficient, the sum of squares $SS = r_{yy}^2 = 1$, and the degree of freedom $C_1 = n - 1$;
- for regression: $C_T = 1$, $SS = r_{iy}^2$;
- for the residual:

$$C_T = n - 2,\ SS = r_{yy}^2 - r_{iy}^2,\ M_s = \frac{SS}{(n-2)}. \tag{7.14}$$

We calculate the next F-criterion:

$$F_T = \frac{M_{S.reg.}}{M_{S.res.}}, \tag{7.15}$$

where:

$M_{S.reg.}$—the mean square due to regression,
$M_{S.res.}$—the mean square due to the residue.
If $F_T < F_{const}$, then the factor X_L is included in the regression.

The inclusion of an influencing factor in the model at any stage of calculation is carried out using the formula:

$$\{\phi \cdot V_{\max}\}/d_{\max} - V_{\max} \cdot 3, \tag{7.16}$$

where: d_{mm}—is the element of the table that allows replacing S at any stage, ϕ—is the number of residual degrees of freedom after including the factor.

Since the factor X_L is included in the regression, the correlation matrix must be transformed. To do this, the L-th row of the matrix A is divided by A_{LL} and a second table is compiled, the elements of which we denote by. They are determined as follows:

$$B_{L1} = \frac{A_{L1}}{A_{LL}}, B_{L2} = \frac{A_{L2}}{A_{LL}}. \tag{7.17}$$

The rest of the table elements are calculated in the following sequence:

$$B_{ij} = A_{ij} - \frac{A_n \cdot A_{ij}}{A_{IJ}}. \tag{7.18}$$

The share of explained variation is determined by the equation:

$$D = M \cdot S_{reg} \cdot 100\%. \tag{7.19}$$

The standard deviation is equal to:

$$S = \sqrt{M \cdot S_{res}}. \tag{7.20}$$

The standardized B—coefficient for X_L is equal to:

$$B_i = \frac{\sum X_i Y_i - (\sum X_i)(\sum Y_i)/n}{\sum X_1^2 - (\sum X_i)^2/n}. \tag{7.21}$$

B—the coefficient for X_L in the original scale is determined from the expression:

$$B_L = B_{Lm} \frac{S_m}{S_L}. \tag{7.22}$$

The standard error of the natural B—coefficient for X_L is equal to:

$$S_{ouu} = \frac{\sqrt{B_L}}{\sum X_L^2}. \tag{7.23}$$

Using partial correlation coefficients, as before, we select as the next factor to be included in the equation such a variable X, which is characterized by the highest partial correlation coefficient with the response.

We determine V_{max} for factors not included in the equation:

$$V_m = \frac{B_{im} \cdot B_{mi}}{B_{ij}}, \quad (7.24)$$

where: $j \neq L, j = 1 \div M$.

To include the factor X, we determine the value of F using Fisher's criterion. If $F > F_{ij}$, then the factor X_j is not included in the regression equation and the use of the stepwise method is concluded.

The constant term of the equation at each step is calculated using the following formula:

$$B_0 = \bar{Y} - \sum_{t=1}^{k} B_i X_L. \quad (7.25)$$

If $F < Fn$ the n-th factor is included in the regression equation, and recalculations are carried out, starting from (7.15).

Since factors Xj and XL are included in the regression equation, partial F—criteria are determined.

The partial F—criterion for X_L, in the presence of X_j in the model, is determined by the formula:

$$F_{xL} = \frac{\phi(C_{Lm} + 1)^2}{(C_{m+1}, C_{m+1})(C_{m+2}, C_{m+2})}, \quad (7.26)$$

where:

ϕ—is the residual number of degrees of freedom.

F—criterion for X_j in the presence of X_L, is equal to:

$$F_{jL} = \frac{\phi(C_{im} + 1)^2}{(C_{m+1}, C_{m+1})(C_{m+2}, C_{m+2})}. \quad (7.27)$$

For factors j and L, the coefficients of variation, deviation of residuals, standardized b-coefficients for X_j and X_L are determined according to the formulas specified in stages (7.18)–(7.20).

If the F-criterion for either the j-th or the L-th factor is less than the critical value, that is, if $F_{xj} < F_{cr}$, then the j-th factor is excluded from the regression equation. In this case, the calculation is redone using expression (7.17).

When calculating V_{max}, the value of F for including a factor is determined. If any k-th factor was just excluded from the regression dependency in the previous step and F for inclusion is greater than the critical F, that is, $F_k > F_{cr}$, then the process of including and excluding factors from the model is stopped, and the free term of the equation is determined by Eq. (7.2).

The mathematical model of SEPC, depending on the obtained x-factors, can be expressed by the following equation:

$$e = B_0 \pm B_1 X_1 \pm \cdots \pm B_k X_k, \tag{7.28}$$

where:

$B_1 \div B_k = 1 \div k$ are natural coefficients for the included factors,
k—numbers of factors included in the model.

The adequacy of the obtained model is determined by substituting the values of the included factors into the regression equation:

$$Z = e - \widehat{e}, \tag{7.29}$$

where:

Z—the residual of the model,
e—original data included for processing,
$\widehat{e}$—the calculated value of e, determined after including influencing factors in the model.

The normal deviation is determined by the following formula:

$$N = \frac{e - \widehat{e}}{\varphi} = \frac{Z}{\varphi}. \tag{7.30}$$

Considering the individual influence of the i-th factor on SEPC, the regression model takes the following form:

$$e = B_0 \pm \sum_{J}^{k} B_J X_J. \tag{7.31}$$

For constant influencing factors, considering the nature of the consumer, we determine the constant value:

$$C = \sum_{j}^{k=1} B_j X_j = const, \tag{7.32}$$

where: k—the number of factors included in the model.

Expression (7.31) is transformed as follows:

$$e = B_0 \pm C \pm B_1 X_1, \tag{7.33}$$

where: $i = 1 \div k$.

The percentage (β^0) change of the i-th factor is determined by the following formula:

$$\beta^0 = X_i \pm \frac{BX_i}{100}. \tag{7.34}$$

The SEPC will change by β percent, which is determined by the following formula:

$$\beta' = \left(\frac{\beta^0 \beta_j + (B_0 \pm C)}{e} - 1 \right) \cdot 100\%. \tag{7.35}$$

Based on the proposed method, factors influencing the SEPC in the steelmaking process of a metallurgical enterprise were identified.

Using the calculated-experimental data and operational information for the period from 2017 to 2020, a statistical analysis of 874 melts was conducted to determine the influence of factors on SEPC. Due to the instability of some data, 598 melts were selected for further statistical processing.

As a result of the analysis, the following factors were considered:

F_1—furnace loading, tons,
F_2—volume of loaded metal, tons,
F_3—volume of loaded metal charge, tons,
F_4—duration of metal melting, min.,
F_5—duration of metal melting under current, min.,
F_6—duration of furnace downtimes, min.,
F_7—mass of the melt, tons.

It follows from [12, 13] that these factors are primary in all metallurgical enterprises, particularly in electrical steelmaking. However, to enhance the accuracy of the calculation, it is necessary to identify the most significant factors. The processing of research results was performed using the method of multiple regression.

Table 7.1 presents the results of the correlation matrix of factors influencing the SEPC during melting. The obtained correlation matrix of influencing factors shows that the SEPC mainly depends on the following three factors: F5—duration of operation under current, minutes (correlation coefficient 0.921), F6—duration of furnace downtimes, minutes (correlation coefficient 0.921), F7—mass of the melt, tons (correlation coefficient 1).

These factors are directly dependent on the SEPC. From the values of the correlation coefficients, it can be concluded that SEPC is strongly dependent on the mass of the melt.

Table 7.1 Correlation matrix of factors

Factors	e	F_1	F_2	F_3	F_4	F_5	F_6	F_7
e	1.000							
F_1	0.214	1.000						
F_2	0.137	0.814	1.000					
F_3	0.459	0.371	0.824	1.000				
F_4	0.657	0.116	0.164	0.224	1.000			
F_5	0.801	0.216	0.294	0.198	0.364	1.000		
F_6	0.921	0.217	0.264	0.149	0.421	0.999	1.000	
F_7	1.000	0.234	0.193	0.193	0.268	0.814	0.899	1.000

The influence of the other factors (F_1, F_2, F_3, F_4) is insignificant and unstable; their correlation coefficients range between 0.02–0.47. The influence of the average scrap share in the primary charge (F_7) on SEPC is inverse (correlation coefficient—0.02).

The SEPC depends on the main influencing factors and is described, according to the mathematical model, by the expression:

$$e = B_0 + B_1F_1 + B_2F_2 + B_3F_3, \tag{7.36}$$

where: $F1$—duration of operation under current, $F2$—duration of furnace downtimes, $F3$—mass of the melt.

B_0, B_1, B_2, B_3—coefficients of the regression equation.

In expression (7.36), the coefficients of the regression equation B_0, B_1, B_2, B_3 remain unknown. These coefficients are determined using the least squares method; the mathematical model (7.36) then takes the form:

$$e = 257965,576 + 2838,082F_1 - 405,456F_2 + 1621,012F_3,$$

The adequacy of the model is checked by Fisher's criterion.

$$F_{cal} > F_{table} : 594 > 434$$

Fulfilling this condition indicates a high degree of adequacy of the developed model. It can be used to determine the calculated values of SEPC considering the influencing factors.

With (7.36), the calculated values of SEPC for steel melting are determined. Table 7.2 presents a comparative analysis of actual and calculated data, which are graphically shown in Fig. 7.1.

Table 7.2 Actual and calculated values of SEPC for steel melting

Sample, *n*	SEPC (actual), thousand kWh/t	SEPC (modeled.), thousand kWh/t	Prediction error	
			Absolute, kW·h	Relative, %
1	460	467	7	1,5
2	493	488	−5	−1,0
3	476	460	−16	−3,4
4	488	495	7	1,4
5	465	470	5	1,1
6	485	480	−5	−1,0
7	493	490	−3	−0,6
8	476	480	4	0,8
9	498	500	2	0,4
10	491	493	2	0,4

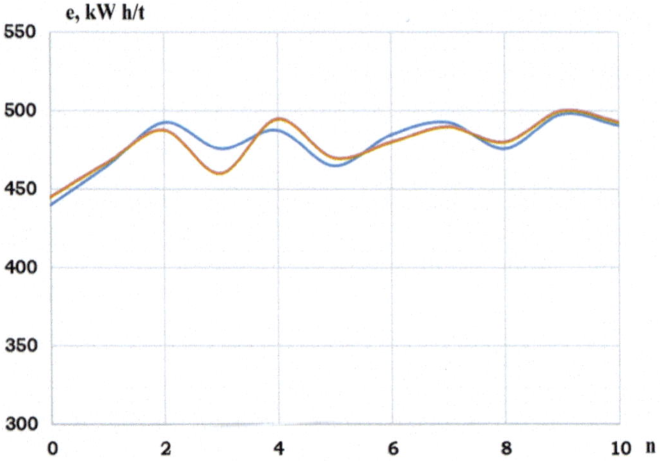

Fig. 7.1 Graphs of actual and calculated SEPC for steel melting

Analysis of Fig. 7.2 shows that the average deviation between actual and calculated values is within 1.9%, indicating the high adequacy of the developed mathematical model for determining SEPC in steel melting. Therefore, it can be applied to determine calculated values of SEPC, particularly in electrical steelmaking.

The proposed method allows for the identification of the most significant factors from the variety influencing energy indicators and provides an appropriate assessment for each. It is unified for all the above-mentioned calculation variants and is included in the corresponding resultant calculations of the norms for SEPC.

In determining the SEPC indicator using the regression-correlation method, factors exerting the greatest influence were identified. The results of the calculations showed that accounting for multiple influencing factors could increase the error in determining the SEPC indicator.

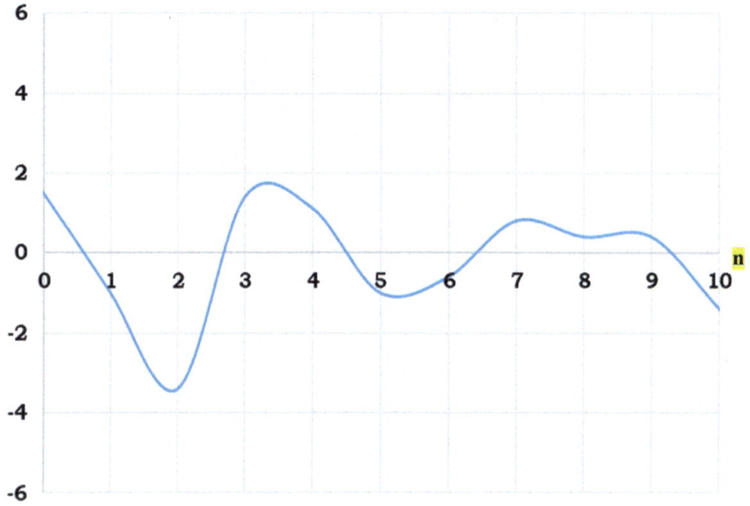

Fig. 7.2 Graph of the discrepancy between actual and calculated values of SEPC in steel melting

In the context of manufacturing processes in workshops and at an industrial enterprise as a whole, the SEPC is a primary technical–economic indicator not only for the unit of product but also for the enterprise's energy management. In the process of designing industrial enterprises, progressive SEPC norms are determined through a reasoned assessment and calculation of this indicator using methods of mathematical statistics and probability theory; they can also be used to determine forecast indicators of electrical power consumption. It should be noted that in forecasting the EPC by most industrial enterprises, the main indicator is its special power consumption [18–20].

As noted earlier, the methods of rationing currently used at industrial enterprises are outdated, and the indicators they establish are unjustified, as they insufficiently consider energy, technological, and production factors. This, in turn, leads to the fact that the error in determining forecast indicators by these methods is higher compared to determining real indicators.

7.3 Methodology for Determining Normative Values of Electrical Power Consumption Based on Multifactor Analysis

At all industrial enterprises, the production of a specific product unit is carried out in several stages. Since a final manufacture is not produced at every stage of the technological process, it is not possible to determine a normative indicator applicable to the stages of manufacture of a unit of product. For this reason, the average value of aggregate indicators of EPC norms is determined based on established normative indicators of different stages of the technological process. It should be noted that the determination of aggregate indicators instead of differentiated ones leads to errors. However, for forecasting purposes, depending on the overall workshop and plant-wide normative indicator, as well as in terms of the technological process, it allows for the control of deviations of normative indicators from permissible values [21, 22].

In this context, an advanced methodology for determining normative indicators of electrical power consumption is proposed. This methodology has been developed using one of the classical methods for solving such tasks—the method of least squares [23, 24]. As noted earlier, there is a correlation between the SEPC indicator and product output; utilizing this correlation allows for the determination of the SEPC indicator by types of products manufactured.

The application of the least squares method in this task enables the determination of unknown coefficients with high precision.

$$\begin{aligned}
&x_1 a_1 + y_1 a_2 + z_1 a_3 + u_1 a_4 + k_1 a_5 + \ldots + W_0 = W \\
&x_2 a_1 + y_2 a_2 + z_2 a_3 + u_2 a_4 + k_2 a_5 + \ldots + W_0 = W \\
&\ldots \\
&x_n a_1 + y_{100} a_2 + z_{100} a_3 + u_{100} a_4 + k_{100} a_5 + \ldots + W_0 = W,
\end{aligned} \tag{7.37}$$

where: x, y, z, u, k—volume of manufactured products; W_0—constant components of EPC in the production of finished goods (ventilation, lighting, etc.); W—the consumption of electricity in the production of finished goods.

It is known that when determining unknown coefficients using a system of normal equations, it is not always possible to ascertain their precise values. In such cases, deviations in the form of $A_i \pm h$ are observed. Equation (7.37) from [25] describes these deviations:

$$\begin{aligned}
&x_i \overline{a}_1 + y_i \overline{a}_2 + z_i \overline{a}_3 + u_i \overline{a}_4 + k_i \overline{a}_5 + W_0 - W = h_i \\
&(i = 1,\ 2,\ 3, \ldots,\ m).
\end{aligned} \tag{7.38}$$

It is established that for determining unknown coefficients, the sum of the squares of the right-hand sides of the Eq. (7.38) must assume a minimum value [26]. By fulfilling this requirement, in solving the task, we minimize the expression (7.38):

$$\begin{aligned}
h_1^2 &= (x_1 \overline{a}_1 + y_1 \overline{a}_2 + z_1 \overline{a}_3 + u_1 \overline{a}_4 + k_1 \overline{a}_5 + \ldots + W_0 - W)^2 \\
h_2^2 &= (x_2 \overline{a}_1 + y_2 \overline{a}_2 + z_2 \overline{a}_3 + u_2 \overline{a}_4 + k_2 \overline{a}_5 + \ldots + W_0 - W)^2 \\
h_3^2 &= (x_3 \overline{a}_1 + y_3 \overline{a}_2 + z_3 \overline{a}_3 + u_3 \overline{a}_4 + k_3 \overline{a}_5 + \ldots + W_0 - W)^2 \\
h_4^2 &= (x_4 \overline{a}_1 + y_4 \overline{a}_2 + z_4 \overline{a}_3 + u_4 \overline{a}_4 + k_4 \overline{a}_5 + \ldots + W_0 - W)^2.
\end{aligned} \tag{7.39}$$

After setting the partial derivatives of the unknown coefficients in (7.39) to zero, the following system of equations is formed [27]:

$$2\sum_{i=1}^{n}(x_i\overline{a}_1 + y_i\overline{a}_2 + z_i\overline{a}_3 + u_i\overline{a}_4 + k_i\overline{a}_5 + \ldots + W_0 - W)x_i = 0;$$

$$2\sum_{i=1}^{n}(x_i\overline{a}_1 + y_i\overline{a}_2 + z_i\overline{a}_3 + u_i\overline{a}_4 + k_i\overline{a}_5 + \ldots + W_0 - W)y_i = 0;$$

$$2\sum_{i=1}^{n}(x_i\overline{a}_1 + y_i\overline{a}_2 + z_i\overline{a}_3 + u_i\overline{a}_4 + k_i\overline{a}_5 + \ldots + W_0 - W)z_i = 0;$$

$$2\sum_{i=1}^{n}(x_i\overline{a}_1 + y_i\overline{a}_2 + z_i\overline{a}_3 + u_i\overline{a}_4 + k_i\overline{a}_5 + \ldots + W_0 - W)u_i = 0;$$

$$2\sum_{i=1}^{n}(x_i\overline{a}_1 + y_i\overline{a}_2 + z_i\overline{a}_3 + u_i\overline{a}_4 + k_i\overline{a}_5 + \ldots + W_0 - W)k_i = 0. \quad (7.40)$$

The values $\overline{a}_1, \overline{a}_2, \overline{a}_3, \overline{a}_4, \overline{a}_5$ are grouped as follows:

$$\left(\sum_{i=1}^{n} x_i^2\right)\overline{a_1} + \left(\sum_{i=1}^{n} x_i y_i\right)\overline{a_2} + \left(\sum_{i=1}^{n} x_i z_i\right)\overline{a_3} + \left(\sum_{i=1}^{n} x_i u_i\right)\overline{a_4} + \left(\sum_{i=1}^{n} x_i k_i\right)\overline{a_5}$$
$$+ \ldots + \sum_{i=1}^{n} x_i W_0 - \sum_{i=1}^{n} x_i W_i = 0;$$

$$\left(\sum_{i=1}^{n} y_i x_i\right)\overline{a_1} + \left(\sum_{i=1}^{n} y_i^2\right)\overline{a_2} + \left(\sum_{i=1}^{n} y_i z_i\right)\overline{a_3} + \left(\sum_{i=1}^{n} y_i u_i\right)\overline{a_4} + \left(\sum_{i=1}^{n} y_i k_i\right)\overline{a_5}$$
$$+ \ldots + \sum_{i=1}^{n} y_i W_0 - \sum_{i=1}^{n} y_i W_i = 0;$$

$$\left(\sum_{i=1}^{n} z_i x_i\right)\overline{a_1} + \left(\sum_{i=1}^{n} z_i y_i\right)\overline{a_2} + \left(\sum_{i=1}^{n} z_i^2\right)\overline{a_3} + \left(\sum_{i=1}^{n} z_i u_i\right)\overline{a_4} + \left(\sum_{i=1}^{n} z_i k_i\right)\overline{a_5}$$
$$+ \ldots + \sum_{i=1}^{n} z_i W_0 - \sum_{i=1}^{n} z_i W_i = 0;$$

$$\left(\sum_{i=1}^{n} u_i x_i\right)\overline{a_1} + \left(\sum_{i=1}^{n} u_i y_i\right)\overline{a_2} + \left(\sum_{i=1}^{n} u_i z_i\right)\overline{a_3} + \left(\sum_{i=1}^{n} u_i^2\right)\overline{a_4} + \left(\sum_{i=1}^{n} u_i k_i\right)\overline{a_5}$$
$$+ \ldots + \sum_{i=1}^{n} z_i W_0 - \sum_{i=1}^{n} z_i W_i = 0;$$

$$\left(\sum_{i=1}^{n} k_i x_i\right)\overline{a_1} + \left(\sum_{i=1}^{n} k_i y_i\right)\overline{a_2} + \left(\sum_{i=1}^{n} k_i z_i\right)\overline{a_3} + \left(\sum_{i=1}^{n} k_i u_i\right)\overline{a_4} + \left(\sum_{i=1}^{n} k_i^2\right)\overline{a_5}$$
$$+ \ldots + \sum_{i=1}^{n} z_i W_0 - \sum_{i=1}^{n} z_i W_i = 0. \quad (7.41)$$

The system of equations (7.41) constitutes a system of normal equations, which are used to determine the desired coefficients [28, 29]:

$$
\begin{gathered}
[x_1x] = \sum_{i=1}^{n} x_i^2; \rightleftarrows\rightleftarrows\rightleftarrows [x_1y] = \sum_{i=1}^{n} x_iy_i; \rightleftarrows\rightleftarrows\rightleftarrows [y_1z] = \sum_{i=1}^{n} y_iz_i; \\
[y_1y] = \sum_{i=1}^{n} y_i^2; \rightleftarrows\rightleftarrows\rightleftarrows [x_1z] = \sum_{i=1}^{n} x_1z_1; \rightleftarrows\rightleftarrows\rightleftarrows [y_1u] = \sum_{i=1}^{n} y_iu_i; \\
[z_1z] = \sum_{i=1}^{n} z_i^2; \rightleftarrows\rightleftarrows\rightleftarrows [x_1u] = \sum_{i=1}^{n} x_1u_1; \rightleftarrows\rightleftarrows\rightleftarrows [y_1k] = \sum_{i=1}^{n} y_ik_i; \\
[u_1u] = \sum_{i=1}^{n} u_i^2; \rightleftarrows\rightleftarrows\rightleftarrows [x_1k] = \sum_{i=1}^{n} x_ik_i; \rightleftarrows\rightleftarrows\rightleftarrows [y_1W_0] = \sum_{i=1}^{n} y_iW_{0i}; \\
[k_1k] = \sum_{i=1}^{n} k_i^2; \rightleftarrows\rightleftarrows\rightleftarrows [x_1W_0] = \sum_{i=1}^{n} x_iW_{0i}; \rightleftarrows\rightleftarrows\rightleftarrows [y_1W] = \sum_{i=1}^{n} y_iW; \\
[W_1W] = \sum_{i=1}^{n} W_i^2; \rightleftarrows\rightleftarrows\rightleftarrows [x_1W] = \sum_{i=1}^{n} x_iW_{0i}; \rightleftarrows\rightleftarrows\rightleftarrows [z_1u] = \sum_{i=1}^{n} z_iu_i; \\
[z_1k] = \sum_{i=1}^{n} z_ik_i; \rightleftarrows\rightleftarrows\rightleftarrows [z_1W_0] = \sum_{i=1}^{n} z_iW_{0i}; \rightleftarrows\rightleftarrows\rightleftarrows [z_1W] = \sum_{i=1}^{n} z_iW_i; \\
[k_1u] = \sum_{i=1}^{n} k_iu_i; \rightleftarrows\rightleftarrows\rightleftarrows [k_1W_0] = \sum_{i=1}^{n} k_iW_{0i}; \rightleftarrows\rightleftarrows\rightleftarrows [k_1W] = \sum_{i=1}^{n} k_iW_i. \ldots
\end{gathered}
\tag{7.42}
$$

After introducing corresponding changes, expressions (7.42) are transformed into the following form [30]:

$$
\begin{aligned}
&[x_1x]\overline{a_1} + [x_1y]\overline{a_2} + [x_1z]\overline{a_3} + [x_1u]\overline{a_4} + [x_1k]\overline{a_5} + x_1W_0 = [x_1W]; \\
&[y_1x]\overline{a_1} + [y_1y]\overline{a_2} + [y_1z]\overline{a_3} + [y_1u]\overline{a_4} + [y_1k]\overline{a_5} + y_1W_0 = [y_1W]; \\
&[z_1x]\overline{a_1} + [z_1y]\overline{a_2} + [z_1z]\overline{a_3} + [z_1u]\overline{a_4} + [z_1k]\overline{a_5} + z_1W_0 = [z_1W]; \\
&[u_1x]\overline{a_1} + [u_1y]\overline{a_2} + [u_1z]\overline{a_3} + [u_1u]\overline{a_4} + [u_1k]\overline{a_5} + u_1W_0 = [u_1W]; \\
&[k_1x]\overline{a_1} + [k_1y]\overline{a_2} + [k_1z]\overline{a_3} + [k_1u]\overline{a_4} + [k_1k]\overline{a_5} + k_1W_0 = [u_1W].
\end{aligned}
\tag{7.43}
$$

This system allows for the determination of unknown coefficients [30].

The smaller the difference between calculated and actual SEPC indicators in the production of goods, the more efficient the EPC in the technological process at the given enterprise. However, it should be remembered that calculated indicators are based on precise calculations. Therefore, it is impractical to take the value of SEPC as a calculated normative indicator and to perform calculations based on this indicator [29].

If the above-mentioned constant components of energy consumption (lighting, compressed air, ventilation, etc.) are taken into account when determining SEPC, it is also essential to consider factors influencing EPC, which enhances the accuracy of the calculations.

7.4 Methodology for Determining Normative Values of EPC Based on Criteria for Minimizing Production Costs

It is acknowledged that an important direction in enhancing the efficiency of energy resource utilization is the optimization of technological production processes. This involves a certain quality criterion of management within a specific period, achieving corresponding constraints from its minimum or maximum values [31].

Enterprises, including their energy services, are classified as complex production systems comprising interconnected subsystems unified by a common operational goal and characterized by a large number of indicators that describe their state [32].

In ultimately increasing production efficiency and reducing costs, the multi-criteria nature of the technological processes of enterprises naturally leads to the pursuit of technological, technical, and organizational solutions that would facilitate targeted changes in multiple indicators simultaneously [33].

Modern equipment in production areas allows for the manufacture of a wide spectrum of high-quality products and is an integral part of several technological processes. Consequently, there is a logical intensification of requirements for the quality of the final products. It is evident that low utilization rates of high-performance equipment, producing high-quality products, can negate all anticipated benefits from the implementation of new technologies.

Typically, the revenues of enterprises (I) consist of two main components: income from the main production (I_{main}) and income from auxiliary productions (I_{aux}):

$$I = I_{main} + I_{aux}. \quad (7.44)$$

Given that industrial enterprises aim to maximize their revenues (achieving I_{max}), one of the tasks addressed in this monograph is to determine the optimal ratio between main and auxiliary production that ensures the attainment of the maximum possible value of expression (7.44) and minimization of production costs.

By employing nonlinear programming methods [34] and constructing a Lagrange function, it is possible to maximize a function of two variables, defined as follows:

$$L = I + \lambda W, \quad (7.45)$$

where: λ—is an indefinite factor, and W—an auxiliary Lagrange function.

As the auxiliary Lagrange function, consider the difference between the annual equivalent costs Z and their limiting value Z_{lim}, which is a finite quantity; this is due to the fundamental limitation of all types of resources:

$$W = Z - Z_{lim} \tag{7.46}$$

Exploring the nature of the equivalent costs for operating technical means of main (Z_{main}) and auxiliary (Z_{aux}) productions:

$$Z = Z_{main} - Z_{aux} = k_n F_{main} + R_{main} + k_n F_{main} + R_{aux}, \tag{7.47}$$

where:

F—the main funds (capital expenditures), comprising,
F_{main}—technical means of the main production,
F_{aux}—technical means of the auxiliary production,
R—annual operational expenses,
R_{main}—expenses for operating the main production technology,
R_{main}—expenses for operating the auxiliary production,
K_n—normative coefficient of economic efficiency.

In expanded form, the Lagrange function (7.45) appears as:

$$\begin{aligned} I &= I_{main} + I_{aux} + \lambda(k_n F_{main} + k_n F_{aux} + R_{main} + R_{aux} - Z_{\text{lim}}) \\ &= I_{main} + I_{aux} + \lambda(F^n_{main} + F^n_{aux} + R_{main} + R_{auc} - Z_{\text{lim}}), \end{aligned} \tag{7.48}$$

where: the terms $F^n_{main} = K_n \cdot F_{main}$ and $F^n_{aux} = K_n \cdot F_{aux}$ are called normalized funds.

To determine the condition for maximizing joint revenues, it is necessary to take partial derivatives of expression (7.48) and set them to zero:

$$\begin{cases} \dfrac{\partial L}{\partial I_{осн}} = 1 + \lambda\left(\dfrac{\partial F^{н}_{осн}}{\partial I_{осн}} + \dfrac{\partial R_{осн}}{\partial I_{осн}}\right) = 0 \\ \dfrac{\partial L}{\partial I_{всп}} = 1 + \lambda\left(\dfrac{\partial F^{н}_{всп}}{\partial I_{всп}} + \dfrac{\partial R_{всп}}{\partial I_{всп}}\right) = 0 \end{cases}. \tag{7.49}$$

Solving the system (7.49) relative to the Lagrange multiplier, we obtain:

$$\lambda = \frac{-1}{\dfrac{\partial F^{н}_{осн}}{\partial I_{осн}} + \dfrac{\partial R_{осн}}{\partial I_{осн}}} = \frac{-1}{\dfrac{\partial F^{н}_{всп}}{\partial I_{всп}} + \dfrac{\partial R_{всп}}{\partial I_{всп}}}. \tag{7.50}$$

Using the equality of denominators, we obtain:

$$\frac{\partial F^{н}_{осн}}{\partial I_{осн}} + \frac{\partial R_{осн}}{\partial I_{осн}} = \frac{\partial F^{н}_{всп}}{\partial I_{всп}} + \frac{\partial R_{всп}}{\partial I_{всп}}. \tag{7.51}$$

Expression (7.51) allows for two equivalent transformations, each with its own economic significance. Let is consider them.

Firstly, consider that equal efficiency of main and auxiliary production is the key to the high efficiency of joint production [35].

Separately group the normalized funds and costs:

$$\frac{\partial F^{н}_{осн}}{\partial I_{осн}} + \frac{\partial F^{н}_{всп}}{\partial I_{всп}} = \frac{\partial R_{осн}}{\partial I_{осн}} + \frac{\partial R_{всп}}{\partial I_{всп}}. \tag{7.52}$$

A consequence of market manipulation of resources in the process of achieving minimal production costs is the economic interpretation of expression (7.52): the difference in increments (with the opposite sign) of annual operational expenses [36].

Secondly, let is return to the representation of the annual equivalent costs in a summarized form (7.47). In this context, expression (7.51) simplifies to:

$$\frac{\partial Z_{осн}}{\partial I_{осн}} = \frac{\partial Z_{всп}}{\partial I_{всп}}. \tag{7.53}$$

It is evident that the converse is also true:

$$\frac{\partial I_{осн}}{\partial Z_{всп}} = \frac{\partial I_{всп}}{\partial Z_{всп}}. \tag{7.54}$$

The expression derived above represents a clear economic interpretation of the increase in revenue per unit of invested resources. The increments in costs for both the main and auxiliary production of the final product should be identical. This conclusion aligns with a fundamental position of economic theory [35]—the condition for achieving maximum revenues, as well as maximum profits for enterprises.

Clearly, well-organized cooperation with partnered producers can be highly attractive—no less so than own production. Typically, in a competitive market, the energy resources of production should be considered as a key component. This is facilitated by the widespread procedure of resource manipulation [36], ensuring the achievement of the lowest costs and maximum profits.

References

1. Martin JE, Rodriguez S (2021) Towards standardization of fuel and energy resource expenditures in contemporary industrial enterprises. Ind Energy J 34(1):11–13
2. Green LA, Thompson RJ (2019) Optimizing energy efficiency in electric steelmaking facilities. J Metall Energy Manage 28(4):22–29
3. Goffman IV (1967) Rationing of energy consumption and energy balances of industrial enterprises. In: Energy, p 319 (in Russian)
4. Martin JE, Liu Y (2022) Analysis of electrical power consumption standardization in the black metallurgy sector. J Ind Energy 40(1):48–53

5. Green S, Brown R (2020) Decision Trees in determining key factors for energy consumption forecasting in steel production. J Metall Innov 35(3):10–13
6. Patel A, Nguyen H (2021) Optimization of energy resources in the textile industry: a contemporary approach. Textile Ind J 29(2):200–210
7. Smith D, Zhang L (2023) New methodologies for electricity standardization in enterprises with dynamic power consumption. Ind Energy Eng 47(4):5–8
8. Johnson E, Garcia M (2021) Calculation of electrical power consumption norms by product type in metallurgical workshops. Electronic J Techn Res 6(1):3–6
9. Thompson R, Lee K (2022) Mathematical modeling of external influences on electrical power consumption parameters. J Sci Inquiry 11(5):37–44
10. White S, Kim J (2023) Analyzing electrical energy expenditure for technological processes in steelmaking facilities. Modern Manage Technol 7(6):143–148
11. Anderson GH, Kumar P (2022) Advanced techniques in correlation-regression analysis of electrical power consumption in the metallurgical industry. Ind Energy Syst Technol 8(4):34–39
12. Hoshimov FA, Rasulov AN, Taslimov AD, Rakhmonov IU (2017) The current state of electrometallurgy in Uzbekistan. Monograph, "East West" Association Studies and Higher Education GmbH. Vienna, Austria, p 84
13. Taylor EJ, Singh RK (2021) Advanced information technology in managing electrical power consumption in steel rolling processes: a case study of global steel manufacturing. Chief Energy Eng J 2:30–35
14. Khoshimov FA, Rakhmonov IU (2017) Energy saving at ferrous metallurgy enterprises. Electronic Sci J "Research of Technical Sciences" 1(19):3–5 (in Russian)
15. Nikiforov GV, Oleynikov VK, Zaslavets BI (2003) Energy saving and energy management in metallurgical production. Energoatomizdat, Moscow, p 480 (in Russian)
16. Johnson LM, Patel S (2022) Analyzing electric power expenditure for technological operations in steel melting shops through correlation-regression methods. J Modern Manage Technol 6(88):154–159
17. Green RA, Thompson EJ (2021) Mathematical modeling of the impact of external factors on electrical power consumption parameters. J Sci Exploration 11(5):14–21
18. Barratt RS (1992) Efficiency of the future. Chem Brit 2:145–148
19. Johnson L, Green R (2022) New methodologies for assessing industrial enterprise development in energy resource utilization. J Ind Efficiency 40(2):10–14
20. Smith D, Zhang L (2021) Advancements in energy consumption management at industrial enterprises. In: Proceedings of the international conference on efficient energy management, pp 25–27
21. Patel A, Nguyen H (2023) Energy management organization for effective electrical power consumption monitoring. J New Inf Technol 32(1):20–25
22. Martin J, Liu Y (2022) Innovative methods for reducing energy expenditures in the textile industry. Textile Ind Energy 38(3):45–50
23. Thompson R, Lee K (2021) Electricity standardization methods for variable power consumption in industrial enterprises. Ind Power Eng 49(6):3–7
24. White S, Kim J (2020) Normative calculations of electrical power consumption in metallurgical industry. Electronic J Techn Res 9(2):2–4
25. Clark E, Brown M (2022) Mathematical modeling of external factors affecting electrical power consumption. J Sci Inquiry 12(5):173–177
26. Garcia M, Wilson R (2023) Correlation-regression analysis of electrical power consumption in metallurgical processes. Modern Manage Technol 8(6):211–216
27. Green S, Brown R (2020) Automation in managing electrical power consumption modes in industrial enterprises. J Precis Mech Control 26(4):42–47
28. Nguyen L, Martin T (2022) Contemporary strategies for energy conservation management in construction. Univ Bull (12):80–85. https://doi.org/10.26428/1827-4278-2022-12-80-85
29. Smith JE, Zhang L (2021) Energy supply management and electrical power consumption: methods and tools. BNTU Press, Minsk, p 210

30. Johnson E, Patel S (2021) Advancements in decision-making methods for industrial electrical power consumption management. Technical University Press, Moscow, p 45
31. Taylor EJ, Singh RK (2023) Modeling and algorithms for industrial electrical power consumption management systems. Kursk State University, p 170. RGB OD, 63:23-5/3466
32. Brown A, Green D (2022) Models and methods for managing electrical power consumption in continuous production industries. Saratov State Technical University, Saratov, p 26
33. White S, Kim J (2021) Management models and algorithms in economic systems for diverse small-batch production. Voronezh Technical University, p 170. RGB OD, 63:21-5/2751
34. Martin J, Liu Y (2022) Automation in electrical power consumption management for machine-building enterprises. Progressive Directions Eng Technol Saratov State Techn Univ 2(14):45–49
35. Rakhmonov IU, Hoshimov FA (2020) Development of an algorithm for evaluating the dominant factors that have the greatest impact on the energy intensity of products. E3S web conference, vol 209. Sustainable Development and Smart Management, ENERGY-21. 07018. https://doi.org/10.1051/e3sconf/202020907018
36. Rakhmonov IU, Tovbaev AN, Nematov LA, Alibekova TSh (2020) Development of forecasted values of specific norms for the issues of produced products in industrial enterprises. J Phys Conf Ser ICMSIT-2020 1515(2020) 022050. https://doi.org/10.1088/1742-6596/1515/2/022050

Chapter 8
Differentiated and Consolidated Specific Norms of Electrical Power Consumption

This chapter discusses the composition and classification of electrical power consumption (EPC) standards (individual, group, technological, and general production). An approximate composition of the norms for electricity consumption by industrial enterprises is given. Consolidated and differentiated standards are described, and the relevance of their application is justified. Methods for determining differentiated and consolidated energy consumption standards are proposed.

8.1 Composition of Norms for the Consumption of Fuel and Energy Resources (FER)

EPC is regulated in conjunction with other energy carriers, primarily thermal energy. Therefore, it is logical to begin the consideration with the regulation of the consumption of FER as a whole, rather than electricity alone.

The high level of modern requirements for the rational use of electrical and thermal energy necessitates a revision of the existing principles of FER consumption regulation. The main directions of these improvements are: expanding the scope of application, ensuring fuller compliance with planning tasks, enhancing a systemic approach, increasing the requirements for the justification of norms, and improving the organization of normative work.

The tasks of regulation should be more fully aligned with planning ones. The primary objective of regulation is to ensure the application in production and planning of technically and economically justified, progressive norms for the consumption of fuel, thermal, and electrical energy, facilitating the regime of FER economy, their most effective use, and rational distribution according to consumption directions.

The objectives of regulation should include the following position—norms for the consumption of fuel, thermal, and electrical energy form the basis for planning the consumption of these resources in production and for evaluating their efficiency.

I. U. Rakhmonov et al., *Electric Consumption by Industrial Enterprises*, Power Systems, https://doi.org/10.1007/978-3-031-62676-0_8

Regulation should cover all fuel, and thermal, and electrical energy expenditures for production and operational needs, regardless of the volume of consumption of these resources and the sources of energy supply.

One of the important principles of FER consumption regulation is the requirement to establish the composition of fuel, thermal, and electrical energy consumption, that is, the items of expenditure included in the norms for production. The composition of fuel and energy consumption norms is established by relevant methodologies and instructions, taking into account production peculiarities. Arbitrary changes to norms are not permitted.

The principle of dimensionality in the regulation of FER consumption is crucial. The dimensions of consumption norms should correspond to the units of measurement used in planning and accounting for fuel, thermal, and electrical energy, volumes of work for product manufacture, and allow for the possibility of monitoring compliance with norms.

The regulation of FER consumption is normalized in the technical system of measurements (calories, kWh, etc.). There is often debate on the necessity of transitioning to a unified international measurement system during FER consumption regulation, but this issue seems premature until it's addressed in planning and accounting. Consumption norms for fuel and energy are developed per unit of production in natural expression, in conditional measurement units, and in monetary terms [1, 2].

The methods used in developing FER consumption norms are of fundamental importance. The scientific validity, progressiveness, systematic nature of the normative planning base, and the timeliness of its development largely depend on the methods used for norm determination.

FER consumption regulation involves establishing a planned measure of consumption. The primary task of regulation is to ensure the application of technically and economically justified progressive consumption norms during planning and production. This is to facilitate a regime of savings, rational distribution, and the most efficient use of resources.

For a comprehensive assessment of the effectiveness of using FER in the production of homogeneous or interchangeable products, generalized specific energy expenditures are determined. These include the direct consumption of all types of energy in the production of a unit of product, converted into units of conditional fuel [3].

Compliance with established consumption norms is a mandatory condition for material incentives for personnel for saving FER. Consumption norms must meet the following requirements:

- consider production conditions, scientific and technical progress, plans for organizational and technical measures aiming at rational and efficient FER use;
- systematically reviewed, taking into account the planned development and technological progress of production, as well as the best indicators of the use of FER;
- contribute to the maximum mobilization of internal reserves for saving electricity and thermal energy, fulfilling planned tasks, and achieving high economic production indicators.

Correctly established norms are one of the important factors stimulating rational use of energy resources; energy savings or increased energy efficiency achieved through organizational and technical measures are reasons for reviewing specific consumption norms.

The implementation of scientifically based progressive energy consumption norms should not be viewed solely from the standpoint of energy resource saving. Their application leads to the emergence of feedback mechanisms: they become a catalyst for technical progress and a factor contributing to the refinement of technological processes and the enhancement of equipment productivity. For this purpose, the norms themselves must be scientifically substantiated and progressive.

The issues of economical energy resource utilization hold national significance and are under continuous supervision by the governments of developed countries. However, in many ministries and agencies, average statistical data of actual expenditures often form the basis for EPC norms and plans. Norms established solely on past statistical data usually fail to consider the potential of technological processes and often do not align with not only optimal but also normal equipment operation modes. Under such norms, their mobilizing role is not just diminished but it is eliminated. With undervalued norms, enterprises lose the incentive to pursue energy savings, as overconsumption will always occur despite all efforts and measures. Energy savings exceeding 5% against established norms indicate their unjustifiably low level. At many enterprises, plans for organizational and technical energy-saving measures are either absent or exist only formally and are not supported by justified technological and economic calculations. The execution of these plans is not monitored, often resulting in their non-realization [4].

Scientifically based norms are founded on the results of comprehensive technological and energy research, in-depth analysis of energy balances (EBs), and critical use of reporting and statistical materials. Special energy indicators, and therefore, the norms, are complex functions of many variable components, possess their industry specificity, and depend on factors characteristic of that particular industrial sector. Their study, analysis, and regulation, in this regard, must be conducted from an industry-specific aspect. Norms need to be dynamic that is, they should be periodically reviewed and improved in line with technical progress, changes in technologies, improvements in product quality, and increased energy armament.

The classification of norms for the consumption of electrical and thermal energy in production is based on the following main characteristics [1]:

- by degree of aggregation: into *individual* and *group* norms;
- by composition of expenditures: into *technological* and *general production* norms;
- by period of validity: into *annual* and *quarterly* norms (monthly norms can also be established at enterprises);
- by scale of application: for *a unit, line, process, workshop, and enterprises.*

An individual norm refers to the SEPC, established for types or individual power-consuming units, installations, machines, technological processes, applicable to specific production conditions.

A group norm is the EPC norm for producing a planned volume of homogeneous products.

A technological norm (workshop, production, factory and plant-based) is the EPC norm accounting for the energy used in primary and auxiliary technological processes for manufacturing a specific type of product. It includes EPC for maintaining technological units in hot standby, for their heating and startup after maintenance and cold downtimes, as well as technically inevitable energy losses during equipment operation.

A general production norm refers to the FER consumption norm that accounts for energy expenditures in primary and auxiliary technological processes, for auxiliary production needs (heating, ventilation, lighting, transport, workshops, etc.), as well as technically inevitable energy losses in converters, thermal and electrical networks, attributed to the production of a given product.

The composition of norms for the consumption of electrical and thermal energy is a list of expenditure items considered in the norms for product production. An approximate composition of these norms for industrial enterprises is provided in Table 8.1. Based on this approximate composition, each enterprise determines the specific composition of consumption norms, taking into account the peculiarities of its product manufacture. Arbitrary changes to the composition of the norms are not allowed [5].

In enterprises manufacturing diverse products, when calculating the standards for the consumption of electrical and thermal energy for auxiliary needs, it is advisable to distribute auxiliary expenses for product manufacturing proportionally to the consumption of energy resources for technological processes of production or, depending on the size of services received from auxiliary and ancillary workshops, namely:

- for transport workshops—proportional to the volume of transported goods;
- for tooling, repair, and other auxiliary workshops—proportional to the share of services;
- for compressor, pumping, or power installations, etc.—proportional to the volumes of compressed air, water, etc. received;
- for the central factory laboratory—proportional to the number of analyses and the volume of experimental work conducted in connection with the production of a given type of product.

Energy losses in electrical, thermal networks, and converters are distributed based on experimental measurements or proportional to the energy consumption in the manufacture corresponding types of products.

Enterprises may establish separate standards for the consumption of electrical and thermal energy for heating, ventilation, lighting, production of compressed air, water supply, and other auxiliary production needs (for which it is necessary to organize the corresponding accounting of energy resources).

The standards for the consumption of electrical and thermal energy in product manufacturing should not include:

Table 8.1 Composition of standards for electrical and thermal energy consumption for industrial enterprises

Types of standards	Articles of thermal and electrical energy consumption included in the standards	Types of energy resources	
		Thermal energy	Electrical energy
Technological	Consumption of electrical and thermal energy for technological processes, considering expenses for maintaining technological units in hot standby, heating them up, and starting after routine repairs and cold downtimes, as well as technically inevitable losses in technological units and installations	+	+
General workshop-level	Consumption of electrical and thermal energy included in the shop-wide technological standard Consumption of thermal and electrical energy for auxiliary needs of the shop		
	• Heating	+	+
	• Ventilation	+	+
	• Lighting	–	+
	• Operation of intra-shop transport	–	+
	• Operation of shop repair workshops	+	+
	• Utilities and sanitary-hygienic needs of the shop (showers, washbasins, personal hygiene rooms for workers)	+	+
	• Energy losses in intra-shop networks and converters		
General production-level (for the entire enterprise)	Consumption of electrical and thermal energy across all productions. Consumption of thermal and electrical energy for auxiliary needs of the enterprise		
	• Water supply	+	+
	• Refrigeration stations	+	+
	• Compressor stations	+	+
	• Central repair and mechanical workshops	+	+
	• Other auxiliary workshops and services (production laboratories, warehouses, administrative buildings, etc.), including their heating, ventilation, and hot water supply	+	+
	• External lighting of the enterprise territory	–	+

(continued)

Table 8.1 (continued)

Types of standards	Articles of thermal and electrical energy consumption included in the standards	Types of energy resources	
		Thermal energy	Electrical energy
	• Transport	–	+
	• Energy losses in thermal and electrical networks and transformers from the location of the enterprise's accounting devices to the points of accounting for productions	+	+

- irrational expenditures and energy losses caused by deviations from the accepted technology, work modes, recipes, as well as costs associated with product defects, malfunctioning equipment, etc.;
- expenses for the construction and major repairs of buildings and structures, installation, commissioning and adjustment of new technological equipment, research and experimental work, supply to external parties, and communal-household needs (canteens, kindergartens and nurseries, clubs, houses of culture, stadiums, rest homes, dormitories, subsidiary farms, etc.), except for objects of communal-household purpose located in the production and administrative buildings of the enterprise, in the absence of separate energy consumption accounting.

8.2 Methods for Determining Differentiated Standards for Electrical Power Consumption

Differentiated operational norms are applicable to specific types and sizes of products manufactured by units of a certain type and capacity. These norms are always established per natural unit of product or per natural unit of completed work volume [1].

If a unit produces a single type of product, the specific operational energy consumption is determined under given operating conditions by a single normalized EB or a single normative energy characteristic, which form the basis for establishing the operational norm.

However, if a unit manufactures products of different types or sizes, the SEPC and their components generally have different values. In other words, each type and size of product should correspond to its own energy characteristic and its own SEPC standard. With a narrow range (assortment) of manufactured products, conducting experiments or calculations necessary to obtain normative energy characteristics for each type or kind of product does not pose significant difficulties and should be considered reasonable, especially for energy-intensive unit operations.

In conditions where products of a more or less wide range are produced on one, and especially on many units, obtaining EBs or characteristics for all types or sizes of products becomes practically impossible or so labor-intensive that it ceases to be justified. In this case, it is more appropriate to establish consolidated norms. But if differentiated norms are still needed under these conditions, their development should be maximally simplified. For this, selective calculations or experiments, or sometimes both, are conducted, and differentiated balances or characteristics are obtained not for all units and types of products, but for individual combinations of both. The choice of the number and objects for experiments or calculations depends on the breadth of the range (assortment) of manufactured products, the number of units involved in their production, and the ratios of energy intensity of different types or sizes of products. Selective experiments or calculations allow establishing not only consolidated norms for groups of products or their entire range but also differentiated norms for those operations that were not covered by the experiments or calculations [6].

Differentiated norms for these operations are established by interpolating data obtained in selective experiments and calculations. The conditions for such interpolation are:

(a) accounting for all manufactured products by a single natural indicator (ton, piece, meter, etc.);
(b) the presence of a range indicator, quantitatively characterizing the differences between one type or size of product from another (diameter, thickness, density, etc.) and reflecting the energy intensity of individual types or sizes of products;
(c) the presence of such a dependency between SEPC and numerical values of the aforementioned indicator, which would be continuous in this range of its variation and would adhere to certain patterns.

It should be noted that not every assortment indicator is capable of reflecting the energy intensity of individual types or sizes of products. This applies, for instance, to the numbers of rolled profiles. Conversely, the yarn number in spinning production can serve as such an indicator, as it numerically reflects the energy intensity of producing yarn of a particular number.

Indicators numerically reflecting the energy intensity of individual types or sizes of products are appropriately termed energy *intensity coefficients*. Energy intensity coefficients can be not only indicators of the product assortment, accepted in production accounting and meeting the requirement of comparing products in terms of energy intensity, but also other indicators, directly considered in production or formed for this purpose based on them. Such indicators may include, for example, paper grammage (in grams per square meter), compression of rolled metal in square centimeters, etc.

It is not always possible to conduct isolated experiments to determine energy expenditures for individual units when manufactured products of various types and sizes on one or several units. Often, production conditions only allow for bulk measurements covering a certain calendar period (shift, day) of manufacture of various types and sizes of products and relating to the joint operation of all or several units (for example, a group of working machines with a common electrical or steam drive, or a group of hydraulic presses operated by a common pumping station). However, it is precisely in these conditions that a purely calculative method of determining norms most often leads to large errors, associated with the difficulty of calculating mechanical losses in working machines and transmitting devices (transmissions, hydraulic networks, etc.).

In the mentioned conditions, the following simple variant of a mixed method, repeatedly tested in practical conditions, can be recommended:

1. During a series of working shifts (or days), while manufacture products of various types and sizes, the total EPC per shift is measured by a group meter. Simultaneously, strict accounting of raw material consumption (semi-finished products, blanks) and output of finished products separately for each type and size is maintained, selective control of working time on individual units is conducted, and their idle times and no-load runs are accounted for.
2. On a non-working day or shift, a series of no-load tests of the group electric motor are conducted, determining the no-load power of the motor without and with transmission. For hydraulic drive units, a no-load test of the piston pumps is performed with the network disconnected and connected, but in its non-working state. Thereby determining the constant losses of the electric motor (electric pump) and constant transmission losses (network leaks).
3. For each experimental working shift, calculate the network for performing the corresponding operations (rolling, drawing, pressing, etc.) with respect to the quantity of manufactured products of each type and size, including the performance of auxiliary operations (trimming ends, moving dies, lifting gates, etc.).
4. The total EPC for each shift is divided by the total network for the shift (expressed in kilowatt-hours), thereby finding the average loss coefficient k_l for each shift:

$$k_l = \frac{W}{W''}, \tag{8.1}$$

   where W—total EPC for the given shift;
   W''—total network for the same shift.
5. Determine the average loss coefficient for all experimental working shifts:

$$k_l = \frac{\sum W_i}{\sum W_i''}. \tag{8.2}$$

If the number of experiments is sufficiently large, the obtained value of the loss coefficient can be considered a reliable average value, corresponding to some average actual operating conditions of the given section.

However, before using this coefficient for the calculation of differentiated norms, it is necessary to adjust it in accordance with possible improvements in operation (reducing the duration of no-load runs of units, reducing the power of constant transmission losses or leaks in the network, etc.).

For this purpose, it is convenient to represent this coefficient in the form of the following product:

$$k_l = k_l' \cdot k_l'', \tag{8.3}$$

where k_l' is the coefficient accounting for losses in electric motors (in electric pumps);

k_l''—is the coefficient accounting for losses associated with the idle operation of working machines, and losses in transmission (leakages in the network). The value of k_l is determined based on passport (specification) data:

(a) for a group electric motor:

$$k_l' = \frac{1}{\eta_m}, \tag{8.4}$$

(b) for electrical pumps:

$$k_l' = \frac{1}{\eta_m \cdot \eta_p}, \tag{8.5}$$

where η_m—is the motor's efficiency at average load,

η_p—is the pump's efficiency.

After this, the coefficient k_l'' is determined:

$$k_l'' = \frac{k_l}{k_l'}. \tag{8.6}$$

The obtained value of k_l'' is analyzed in the context of reducing the mentioned losses. To assess the potential reduction in losses associated with the idle operation of working machines, timekeeping of working hours is used. The assessment of potential reduction in losses in the power supply networks (leakages in the network) is based on the results of no-load tests (see Sect. 8.2) and on data about the loss coefficient k_l for individual days. By adjusting the coefficient k_l'', the corrected value of the total loss coefficient k_l is obtained.

6. By taking the adjusted coefficient of total losses as average for the normal working conditions of the section and multiplying it by the calculated value of the specific network per unit of production for each size type, differentiated norms of SEPC for all size types are obtained.

8.3 Method for Calculating Aggregated Norms of Electrical Power Consumption

Unlike differentiated norms, which apply to individual types and sizes of products or to individual units, aggregated norms are applicable to groups of products of various types and sizes, or to groups of units with different efficiencies and productivity.

The simplest case of formulating aggregated norms relates to the production or transportation of a single product by several units of varying efficiency (for example, annealing a single metal product in furnaces of different types and varying productivity; compression of gas or air in compressors of different productivity) [7]. In this case, the aggregated norm is determined as a weighted average value of EPCs for individual units, in accordance with the shares of production by individual units in the total output. To calculate the aggregated norm in this case, it is necessary to know the distribution of production among the units and have differentiated norms for individual units corresponding to the given load of production output for each unit. If there is a group characteristic of the units, the aggregated norm can be obtained directly from this characteristic. The need to establish an aggregated norm here is determined by the unity of the manufactured product.

In a more general case of producing mixed products on one or several units, the transition from differentiated to aggregated norms is dictated by other reasons [7].

With a wide range of products, it is difficult, and sometimes practically impossible, to cover all combinations of units and product sizes with differentiated norms. In this case, the transition from differentiated to aggregated norms should primarily be considered as a means to simplify the listed types of standardization and, thereby, ensure as complete a coverage as possible by the norms of individual units and product sizes.

Establishing differentiated norms in the context of systematic changes in the product range is no less labor-intensive. In this scenario, using aggregated norms instead of differentiated ones greatly simplifies the planning and control of EPC, although it often leads to additional errors in determining the average EPC per unit of total product output. The transition to aggregated norms is dictated, in addition to the above, also by the difficulties of typifying differentiated norms on an industry scale, their approval by ministries, and their control by inspection bodies. It should be noted that not only operational norms can be aggregated, but also total norms for production processes of any scale (workshop, department, enterprise) [8].

Establishing norms, differentiated by all types and sizes of products, is justified only for the most energy-intensive items, and even then only with a narrow range of products (for example, different types of cast iron in blast furnace production). In the majority of cases, it is more practical to establish and use aggregated norms. In compiling aggregated norms, determining differentiated SEC is indispensable. However, it is not necessary to determine the latter for all types and sizes of products; it can be limited to selective experiments or calculations.

In the context of production processes in workshops and at industrial enterprises as a whole, the SEC is a basic technical–economic indicator not only for the product unit but also for the enterprise's energy management. In the design process of industrial enterprises, progressive SEC norms are determined through reasoned assessment and calculation of this indicator using methods of mathematical statistics and probability theory; they can also be used to determine estimated indicators of EPC [9].

As noted above, in the technological process of industrial enterprises with continuous production, it is impossible to carry out experimental investigations regarding the justification of EPC volumes. Therefore, it is advisable to establish aggregated norms for such enterprises, especially for energy-intensive enterprises in the ferrous metallurgy sector. (These are used as examples in the monograph.)

In metallurgical enterprises, as well as in industrial enterprises of other profiles, the production of a certain unit of product is carried out in several stages. Since a final product is not manufactured at every stage of the technological process, it is not possible to determine a normative indicator applicable to the stages of manufacture of a product unit. For this reason, the average value of aggregated SEC norms is determined based on the established normative indicators of different stages of the technological process. It should be noted that determining aggregated indicators, instead of differentiated ones, leads to errors; however, in forecasting, depending on the overall workshop and overall plant normative indicator, as well as in the context of the technological process, it allows for controlling the deviation of normative indicators from acceptable values [10, 11].

In this regard, the monograph presents an advanced methodology for determining aggregated normative indicators. The methodology is developed using one of the classic methods for solving such tasks, namely the least squares method [12, 13]. Since there is a correlation between the indicator of SEPC and the volume of produced goods, it allows for determining the SEC indicator by types of produced goods.

The least squares method, applied in solving this task, allows for determining unknown coefficients with high precision.

$$\begin{aligned}
&x_1a_1 + y_1a_2 + z_1a_3 + u_1a_4 + k_1a_5 + \cdots + W_0 = W;\\
&x_2a_1 + y_2a_2 + z_2a_3 + u_2a_4 + k_2a_5 + \cdots + W_0 = W;\\
&\quad\ldots \qquad\qquad\qquad\qquad ;\\
&x_{100}a_1 + y_{100}a_2 + z_{100}a_3 + u_{100}a_4 + k_{100}a_5 + \cdots + W_0 = W,
\end{aligned} \tag{8.7}$$

where x, y, z, u, k—the volume of produced goods,

W_0—constant components of EPC in the production of finished goods (ventilation, lighting, etc.),

W—EPC to produce finished goods.

It is known that when determining unknown coefficients using a system of normal equations, it is not always possible to determine their exact values. In such cases, a deviation of the form $A_i \pm h$ is observed. The equation describing the deviations has the form [14]:

$$\begin{aligned}
&x_i\bar{a}_1 + y_i\bar{a}_2 + z_i\bar{a}_3 + u_i\bar{a}_4 + k_i\bar{a}_5 + W_0 - W = h_i\\
&(i = 1, 2, 3, \ldots, m).
\end{aligned} \tag{8.8}$$

It is known that to determine unknown coefficients, the sum of the squares of the right-hand sides of Eq. (8.8) must take a minimum value [15, 16]. Fulfilling this requirement, when solving the task at hand, we minimize expressions (8.9):

$$\begin{aligned} h_1^2 &= (x_1\overline{a}_1 + y_1\overline{a}_2 + z_1\overline{a}_3 + u_1\overline{a}_4 + k_1\overline{a}_5 + \cdots + W_0 - W)^2; \\ h_2^2 &= (x_2\overline{a}_1 + y_2\overline{a}_2 + z_2\overline{a}_3 + u_2\overline{a}_4 + k_2\overline{a}_5 + \cdots + W_0 - W)^2; \\ &\quad \ldots \qquad ; \\ h_4^2 &= (x_{100}\overline{a}_1 + y_{100}\overline{a}_2 + z_{100}\overline{a}_3 + u_{100}\overline{a}_4 + k_{100}\overline{a}_5 + \cdots + W_0 - W)^2 \end{aligned} \tag{8.9}$$

After setting the partial derivatives of the unknown coefficients in (8.9) to zero, the following system of equations is formed [17]:

$$\begin{aligned} &2\sum_{i=1}^{n} (x_i\overline{a}_1 + y_i\overline{a}_2 + z_i\overline{a}_3 + u_i\overline{a}_4 + k_i\overline{a}_5 + \cdots + W_0 - W)x_i = 0; \\ &2\sum_{i=1}^{n} (x_i\overline{a}_1 + y_i\overline{a}_2 + z_i\overline{a}_3 + u_i\overline{a}_4 + k_i\overline{a}_5 + \cdots + W_0 - W)y_i = 0; \\ &2\sum_{i=1}^{n} (x_i\overline{a}_1 + y_i\overline{a}_2 + z_i\overline{a}_3 + u_i\overline{a}_4 + k_i\overline{a}_5 + \cdots + W_0 - W)z_i = 0; \\ &2\sum_{i=1}^{n} (x_i\overline{a}_1 + y_i\overline{a}_2 + z_i\overline{a}_3 + u_i\overline{a}_4 + k_i\overline{a}_5 + \cdots + W_0 - W)u_i = 0; \\ &2\sum_{i=1}^{n} (x_i\overline{a}_1 + y_i\overline{a}_2 + z_i\overline{a}_3 + u_i\overline{a}_4 + k_i\overline{a}_5 + \cdots + W_0 - W)k_i = 0. \end{aligned} \tag{8.10}$$

The values of $\overline{a_1}, \overline{a_2}, \overline{a_3}, \overline{a_4}, \overline{a_5}$, are grouped in the following way: (To be listed in a line.)

$$\begin{aligned} &\left(\sum_{i=1}^{n} x_i^2\right)\overline{a}_1 + \left(\sum_{i=1}^{n} x_i y_i\right)\overline{a}_2 + \left(\sum_{i=1}^{n} x_i z_i\right)\overline{a}_3 + \left(\sum_{i=1}^{n} x_i u_i\right)\overline{a}_4 \\ &\quad + \left(\sum_{i=1}^{n} x_i k_i\right)\overline{a}_5 + \cdots + \sum_{i=1}^{n} x_i W_0 - \sum_{i=1}^{n} x_i W_i = 0; \\ &\left(\sum_{i=1}^{n} y_i x_i\right)\overline{a}_1 + \left(\sum_{i=1}^{n} y_i^2\right)\overline{a}_2 + \left(\sum_{i=1}^{n} y_i z_i\right)\overline{a}_3 + \left(\sum_{i=1}^{n} y_i u_i\right)\overline{a}_4 \\ &\quad + \left(\sum_{i=1}^{n} y_i k_i\right)\overline{a}_5 + \cdots + \sum_{i=1}^{n} y_i W_0 - \sum_{i=1}^{n} y_i W_i = 0; \end{aligned}$$

$$\left(\sum_{i=1}^{n} z_i x_i\right)\overline{a}_1 + \left(\sum_{i=1}^{n} z_i y_i\right)\overline{a}_2 + \left(\sum_{i=1}^{n} z_i^2\right)\overline{a}_3 + \left(\sum_{i=1}^{n} z_i u_i\right)\overline{a}_4$$
$$+ \left(\sum_{i=1}^{n} z_i k_i\right)\overline{a}_5 + \cdots + \sum_{i=1}^{n} z_i W_0 - \sum_{i=1}^{n} z_i W_i = 0;$$
$$\left(\sum_{i=1}^{n} u_i x_i\right)\overline{a}_1 + \left(\sum_{i=1}^{n} u_i y_i\right)\overline{a}_2 + \left(\sum_{i=1}^{n} u_i z_i\right)\overline{a}_3 + \left(\sum_{i=1}^{n} u_i^2\right)\overline{a}_4$$
$$+ \left(\sum_{i=1}^{n} u_i k_i\right)\overline{a}_5 + \cdots + \sum_{i=1}^{n} z_i W_0 - \sum_{i=1}^{n} z_i W_i = 0;$$
$$\left(\sum_{i=1}^{n} k_i x_i\right)\overline{a}_1 + \left(\sum_{i=1}^{n} k_i y_i\right)\overline{a}_2 + \left(\sum_{i=1}^{n} k_i z_i\right)\overline{a}_3 + \left(\sum_{i=1}^{n} k_i u_i\right)\overline{a}_4$$
$$+ \left(\sum_{i=1}^{n} k_i^2\right)\overline{a}_5 + \cdots + \sum_{i=1}^{n} z_i W_0 - \sum_{i=1}^{n} z_i W_i = 0. \tag{8.11}$$

The system of equations (8.11) constitutes a system of normal equations, which are used to determine the desired coefficients [18, 19]:

$$\begin{aligned}
&[x_1 x] = \sum_{i=1}^{n} x_i^2; \rightleftarrows\rightleftarrows\rightleftarrows [x_1 y] = \sum_{i=1}^{n} x_i y_i; \rightleftarrows\rightleftarrows\rightleftarrows [y_1 z] = \sum_{i=1}^{n} y_i z;\\
&[y_1 y] = \sum_{i=1}^{n} y_i^2; \rightleftarrows\rightleftarrows\rightleftarrows [x_1 z] = \sum_{i=1}^{n} x_1 z_1; \rightleftarrows\rightleftarrows\rightleftarrows [y_1 u] = \sum_{i=1}^{n} y_i u_i;\\
&[z_1 z] = \sum_{i=1}^{n} z_i^2; \rightleftarrows\rightleftarrows\rightleftarrows [x_1 u] = \sum_{i=1}^{n} x_1 u_1; \rightleftarrows\rightleftarrows\rightleftarrows [y_1 k] = \sum_{i=1}^{n} y_i k_i;\\
&[u_1 u] = \sum_{i=1}^{n} u_i^2; \rightleftarrows\rightleftarrows\rightleftarrows [x_1 k] = \sum_{i=1}^{n} x_1 k_1; \rightleftarrows\rightleftarrows\rightleftarrows [y_1 W_0] = \sum_{i=1}^{n} y_i W_{0i};\\
&[k_1 k] = \sum_{i=1}^{n} k_i^2; \rightleftarrows\rightleftarrows\rightleftarrows [x_1 W_0] = \sum_{i=1}^{n} x_1 W_{0i}; \rightleftarrows\rightleftarrows\rightleftarrows [y_1 W] = \sum_{i=1}^{n} y_i W;\\
&[W_1 W] = \sum_{i=1}^{n} W_i^2; \rightleftarrows\rightleftarrows\rightleftarrows [x_1 W] = \sum_{i=1}^{n} x_1 W_{0i}; \rightleftarrows\rightleftarrows\rightleftarrows [z_1 u] = \sum_{i=1}^{n} z_i u_i.
\end{aligned} \tag{8.12}$$

After introducing the corresponding changes, expressions (8.12) are transformed into the following form [20, 21]:

$$
\begin{aligned}
&[x_1x]\overline{a_1} + [x_1y]\overline{a_2} + [x_1z]\overline{a_3} + [x_1u]\overline{a_4} + [x_1k]\overline{a_5} + x_1W_0 = [x_1W];\\
&[y_1x]\overline{a_1} + [y_1y]\overline{a_2} + [y_1z]\overline{a_3} + [y_1u]\overline{a_4} + [y_1k]\overline{a_5} + y_1W_0 = [y_1W];\\
&[z_1x]\overline{a_1} + [z_1y]\overline{a_2} + [z_1z]\overline{a_3} + [z_1u]\overline{a_4} + [z_1k]\overline{a_5} + z_1W_0 = [z_1W];\\
&[u_1x]\overline{a_1} + [u_1y]\overline{a_2} + [u_1z]\overline{a_3} + [u_1u]\overline{a_4} + [u_1k]\overline{a_5} + u_1W_0 = [u_1W];\\
&[k_1x]\overline{a_1} + [k_1y]\overline{a_2} + [k_1z]\overline{a_3} + [k_1u]\overline{a_4} + [k_1k]\overline{a_5} + k_1W_0 = [k_1W].
\end{aligned}
\tag{8.13}
$$

This system allows for the determination of the unknown coefficients [21].

When the difference between the calculated and actual indicators of SEPC in the production of goods is reduced, the efficiency of EPC in the technological process at the industrial enterprise increases. It is important to remember that the indicators determined by calculation are based on precise calculations. This eliminates the need to consider the SEC value as a calculated normative indicator and to perform calculations based on this indicator [18, 19].

If the above-mentioned constant components of EPC (lighting, compressed air, ventilation, etc.) are accounted for in determining the SEC, it is also necessary to consider factors affecting EPC, which increases the accuracy of the calculations.

References

1. Goffman IV (1966) Rationing of energy consumption and energy balances of industrial enterprises. Energy 319
2. Oksana V (2020) Consumption of fuel and energy resources at agricultural enterprises. Innov Econ 4(26):65–72. https://doi.org/10.37332/2309-1533.2020.5-6.9
3. Sioshansi FP (2020) Consumer, prosumer, prosumager: how service innovations will disrupt the utility business model. Academic Press
4. Zhao X, Lynch C, Dahl R (2020) Energy efficiency and conservation in smart cities. Energy Policy 139:111335
5. Shenets Ya (2022) Evaluation of the energy efficiency of industrial furnaces based on the modeling of fuel consumption modes. Energy Proc CIS High Educ Inst Power Eng Assoc 65(2):169–180. https://doi.org/10.21122/1029-7448-2022-65-2-169-180
6. Vries D, Wiegmann HJ, Blind P, Mangelsdorf A (2020) Effects of standardization on firm performance: an empirical study. Technol Forecast Soc Change 160:120221
7. Lev K, Tatyana B, Olga K, Alexandra Z (2015) Method of multilevel rationing and optimal forecasting of volumes of electric-energy consumption by an industrial enterprise. Autom Control Comput Sci 48:324–333. https://doi.org/10.3103/S0146411614060054
8. Patel A, Nguyen H (2021) Optimization of energy resources in the textile industry: a contemporary approach. Text Ind J 29(2):200–210
9. Volkov AI, Stulov PE, Leont'ev LI, Uglov VA (2020) Analysis of the use of rare earth metals in ferrous metallurgy of Russia and world. Izv Ferrous Metall 63(6):405–418 (in Russian). https://doi.org/10.17073/0368-0797-2020-6-405-418
10. Gnatyuk VI et al (2020) Optimal control of power consumption of the regional electrical complex using rank analysis methods: educational and methodological manual. Publishing House of the IKBFU. I. Kanta, Kaliningrad, p 233 (in Russian)
11. Khoshimov FA, Rakhmonov IU (2019) Analysis of power consumption regulation at ferrous metallurgy enterprises. J Bull Tashkent State Tech Univ Tashkent 1:52–58 (in Russian)
12. http://www.vevivi.ru/best/Klassicheskii-metod-naimenshikh-kvadratov-ref108272.html

13. Brown JE, Nguyen LK (2022) Advanced application of the least squares method in modeling inelastic deformation. J Samos Tech Univ Ser Phys Math Sci 6:125–128. https://doi.org/10.14498/vsgtu2022
14. Martin RA, Green SL (2021) Mathematical modeling of electrical power consumption by industrial systems. Energy Transp J 3(11):20–26
15. Patel II, Kumar AJ (2023) Multifactorial modeling of electrical power consumption in an industrial enterprise, university reports. Electromechanics (2/3):70–73
16. Taylor VM, Spencer DV (2022) Neural network diagnostic model for computational networks, information and control systems. Railw Transp (5):15–21
17. Johnson AI, Brown GV (2023) Standardizing electrical power consumption through mathematical modeling, university news. Energy Probl 9–10:128–132
18. White GR, Ross ED (2024) Mathematical model for predicting electrical load in housing and communal services: a case study of multi-storey car parks, university reports. Energy Issues 11–12:12–14
19. Samoseiko VF (1995) Mathematical modeling of electricity consumption by production systems. Electricity 3:23–29 (in Russian)
20. Rakhmonov IU, Tovbaev AN, Nematov LA, Alibekova TSh (2020) Development of forecasted values of specific norms for the issues of produced products in industrial enterprises. J Phys Conf Ser ICMSIT-2020 1515:022050. https://doi.org/10.1088/1742-6596/1515/2/022050
21. Rakhmonov IU, Hoshimov FA (2020) Development of an algorithm for evaluating the dominant factors that have the greatest impact on the energy intensity of products. E3S Web Conf 209:07018. Sustainable development and smart management, ENERGY-21. https://doi.org/10.1051/e3sconf/202020907018

Chapter 9
Forecasting Electrical Power Consumption

In the chapter, the primary requirements and stages for developing models to forecast electrical power consumption (EPC) are proposed. To enhance the accuracy of forecast indicators, prediction models utilizing artificial neural network methodology are developed. The advantages and disadvantages of various estimation methods are analyzed; corresponding algorithms and software products for this purpose are proposed.

9.1 Forecasting Models for Electrical Power Consumption

An effective way to prevent conflicts between industrial enterprises and power supply organizations regarding the cost and volume of electricity supplied/consumed is to improve the accuracy of forecasting EPC volumes and schedules. This improvement is achieved by determining these parameters for the contractual accounting period. When determining forecast indicators, it is advisable to use modern high-precision forecasting methods.

Each enterprise should preliminarily determine the magnitude and schedule of power consumption with the highest possible accuracy and then strictly adhere to them, thereby minimizing penalties and fines. Enterprises often pay significant fines for consuming electricity beyond the established limit. This is primarily due to the lack of advanced methods for forecasting EPC at industrial enterprises, i.e., the presence of large errors in forecasts due to the imperfection of the applied methods.

An analysis of forecasting methods used in practice shows that they are predominantly based either on expert assessments of EPC volumes or on actual consumption records. EPC forecasting provides initial information for planning normal operational modes in energy management. Based on the forecast results, the actual and optimal modes of power supply systems are calculated, and the quality parameters of electricity supply (reliability, power quality, etc.) are assessed.

I. U. Rakhmonov et al., *Electric Consumption by Industrial Enterprises*, Power Systems, https://doi.org/10.1007/978-3-031-62676-0_9

Most existing algorithms for forecasting EPC, developed in the energy sector, are a combination of various statistical methods. There are methods in which the load is represented as a random process. There are also forecasting techniques that identify a base (main) component in load changes. However, qualitative modeling is a very complex process due to the nonlinear relationships between factors and the load on which it depends. Furthermore, it should be noted that known methods for forecasting EPC do not work with distorted or incomplete data. In this regard, entirely new methods and approaches for forecasting EPC are necessary, taking into account heterogeneous data.

A number of ideas in the field of estimation EPC indicators, primarily based on linear models, were put forward quite some time ago. For example, in works [1, 2], the analysis relies on statistical analysis methods developed in the early and mid-last century:

- modeling development processes;
- spatial and temporal extrapolation;
- heuristic and econometric;
- factorial, regression, and correlation.

Currently, much attention is paid to methods based on nonlinear models. Most of these methods belong to the class of artificial intelligence technologies. Since the currently used forecasting methods do not consider the nonlinearity of the dependency of EPC on factors affecting forecast indicators, the difference between actual and forecasted EPC indicators often exceeds permissible values. To overcome this obstacle, it is necessary to develop mathematical models that consider the patterns of change in EPC, considering the real nature of the technological process at the enterprise. Moreover, it is necessary to determine the degree of influence of various factors on EPC. For improving the accuracy of the forecast, primary data is important; the smaller the intervals for obtaining this data, the higher the adequacy of the developed models.

9.1.1 Key Requirements for EPC Forecasting Models

It is important to note that existing forecasting methods have evolved over time in conjunction with the development of the production process. Among these methods, it is crucial to select one that is adequate to the characteristics of the research object and reflects all elements of the technological process. Forecasting models are typically created for a day, week, month, and year [3, 4].

The second most important requirement in developing forecasting models is their level of adaptability. This means that the model developed for a particular case should be suitable for use in other enterprises within the same sector. However, it is essential first and foremost to consider the several key factors that most strongly influence the specific object for which the estimation model is developed.

Since EPC can be influenced simultaneously by a number of factors, including technological, production, and energy factors, forecasting models should be based on the patterns of change in EPC.

Forecasting models rely on either *unifactorial* or *multifactorial analysis*, taking into account the number or nature of factors influencing EPC, which in turn depends on the specifics of the industrial enterprise.

Unifactorial analysis is used when the forecasting model is developed considering one factor that most strongly influences EPC.

Multifactorial analysis considers two or more factors. Generally, it is undesirable to deal with a large number of factors in analysis. The reason is that including factors in the model without studying their impact on EPC can increase the estimation error. Determining the level of influence of various factors is a multi-faceted task, for which the regression-correlation method is used. The dependency level of several factors on EPC is determined; factors with high dependency are included in the forecasting model [5, 6].

It is known that an industrial enterprise forms its future plans based on indicators determined by the estimation model. Essentially, the enterprise develops a "roadmap" for itself. Given this, ensuring acceptable forecasting accuracy requires great skill and responsibility from the specialists developing the model. As noted earlier, not only factors but also the composition of the initial data are important in developing forecasting models. In this case, it is advisable to change the initial data used for forecasting by law. The reason is that the data have a highly variable character, leading to the assessment of these improbable changes in the model and, as a consequence, to a decline of accuracy [7, 8].

Stability of the initial data is a requirement that is one of the main ones for the forecasting model and is directly related to meeting the above-mentioned accuracy requirement. When forming a known initial database, the content of this initial data is analyzed before developing forecasting models using it, i.e., from unstable information [1, 9]. Currently, the stability of initial data can be ensured using modern forecasting methods.

9.1.2 Principal Stages in Developing EPC Forecasting Models

The development of EPC forecasting models can be conventionally divided into five stages:

1. *Researching the structure of EPC forecasting.*

Before developing a forecasting model, it is necessary to select the most suitable method from the existing forecasting methods for the object of study and develop a model based on it. Forecasting parameters include periodicity, dependent on the forecasting object and encompassing certain patterns of EPC change. As mentioned earlier, preliminary data, necessary for model development, play a crucial role in enhancing the accuracy of the estimation model. This stage determines the composition of the initial data required for the model [10, 11]. The more primary data included, the more comprehensively the model will account for all changes over the period, thereby reducing model error and increasing adequacy [12–14].

2. *Identifying key and auxiliary factors influencing the accuracy of EPC forecast.*

This ensures increased accuracy of the forecasting model. In general, more than ten factors, related and unrelated to the technological process, can influence EPC. The number of these factors is determined using the expert assessment method [15, 16].

3. *Determining the interrelationships among factors selected for inclusion in the EPC forecasting model.*

This stage is crucial in ensuring high accuracy of the developed model. Not all factors should be included in the forecasting model. Including additional factors or, conversely, excluding some, requires additional calculations. Currently, these calculations are based on the multivariate regression method, and factors are included in the estimation model based on mathematical models expressing the relationship between EPC and influencing factors. The value of correlation coefficients is determined based on these [6].

4. *Studying the impact of changes in EPC dynamics on the forecasting model.*

The volume of production at industrial enterprises, in the context of the modern market economy, is directly managed by market laws. This, in turn, means that changes in market demand for produced goods alter EPC. In such conditions, developing models for forecasting EPC becomes significantly more complex. Usually, the construction of forecasting models requires the use of stable initial data. In this case, a sharp change in the dynamics of EPC causes instability. Therefore, when constructing a forecasting model, it is necessary to include corrective coefficients considering the dynamics of EPC changes [17].

5. *Constructing the EPC forecasting model and identifying and assessing its errors.*

After completing the research conducted in the four previous stages, the forecasting model is constructed. Using the forecasting model, projected indicators of EPC are determined. These projected indicators are compared with the actual indicators corresponding to the considered period, and their errors are identified [18].

It should be noted that forecasting models of an object are developed based on several alternative methods. Naturally, the model ensuring the smallest error is chosen. The above requirements for developing EPC forecasting models are general in nature and should be specified according to the peculiarities of production processes. In particular, failing to account for certain factors influencing estimation accuracy can lead to significant errors in determining projected indicators.

9.2 Advantages and Disadvantages of Applied Electrical Power Consumption Forecasting Methods

The advantages and disadvantages of various methods for forecasting EPC are summarized in Table 9.1 [3, 4].

Analyzing the purpose and tasks of EPC in the electricity market, as well as the advantages and disadvantages of the main methods of forecasting time series, allows us to draw the following conclusions [3, 4]:

- The forecasting model should provide optimal accuracy in terms of cost–benefit ratio, taking into account the costs of the automated software and hardware system and the costs associated with deviations of the predicted EPC from actual consumption;

Table 9.1 Advantages and disadvantages of electrical power consumption estimation methods

Estimation models	Advantages	Disadvantages
Intuitive forecasting methods	Applicability in situations where formalized forecasting on methods are not feasible	Dependence of forecast quality on the expertise of analysts, limited potential for forecast automation
Forecast extrapolation	Ease of implementing the forecasting model, the ability to make predictions based on a minimal set of input data	Rigid fixation of the trend model, lack of adaptability, inability to account for a multitude of influencing parameters, need for a large volume of initial data
Correlation and regression analysis	The ability to consider the influence of a wide range of parameters on EPC	Low adaptability, stringent requirements for initial data, complexity of implementation
ARIMA models	Strong capability to identify and account for seasonal influencing parameters	Low adaptability, complexity of implementation, reduced accuracy due to high uncertainty in EPC volume
Adaptive forecasting methods	Good adaptive abilities, suitable for long-term estimation	Complexity in determining the set of input parameters, necessity for a large set of input data
Forecasting using artificial neural networks (ANN)	Good adaptive and generalizing abilities, high noise immunity, capability for self-learning	Difficulty in determining the set of input parameters and network configuration
Forecasting with hybrid systems based on ANNs	Good adaptive and generalizing abilities, high noise immunity, self-learning and self-tuning capabilities using ANNs, potential for automation	Complexity of implementation

- The forecasting model should be adaptive, capable of quickly altering its properties under the influence of changing external environmental factors;
- Among all the properties, the best methods for developing forecasting techniques as a basis for an automated forecasting model are those based on artificial neural networks (ANNs). These methods are distinguished by the fact that the identification of EPC patterns and the construction of the forecasting model occur simultaneously during the neural network training process;
- To determine the optimal configuration of ANNs, it is necessary to use the method of genetic selection;
- The most suitable for creating an EPC forecasting model is a hybrid system based on a multilayer perceptron, created using genetic algorithms to select optimal network parameters.

9.3 Forecasting Models Based on the Method of Artificial Neural Networks

Choosing EPC forecasting methods that consider the nature of the object contributes to achieving high accuracy in forecast indicators. Currently, artificial intelligence methods are widely used in forecasting parameters of energy sector objects. Using these methods allows for selecting the most accurate parameters based on a multi-stage learning process. In particular, using the ANN method enables highly accurate EPC forecasting [19].

In developing forecasting models, the ANN method has several advantages compared to other methods [20]:

- the capability for training and self-learning until the conditions set during forecasting are met;
- the acceptability of the training process even with changes in the composition and volume of initial data;
- the possibility of obtaining highly accurate forecast indicators with the correct choice of composition and volume of initial data;
- adaptability, allowing its successful application to solve a wide range of tasks in forecasting.

In estimation energy consumption using the ANN method, adjustable local linear autoregressive forecasting is applied:

$$x^{f}_{n+1} = a^{/}_{0}x_n + a^{/}_{1}x_{n-1} + \cdots + a^{/}_{m-1}x_{n-(m-1)} + a^{/}_{m}. \tag{9.1}$$

The coefficients in expression (9.1) are determined by the method of least squares (MLS), a classical method for determining constant coefficients.

In applying MLS, the condition for the best approximation is:

$$\sum_{r=1}^{k}\left(x_{ir+1}-a_0x_{ir-1}-\cdots-a_{m-1}x_{ir-(m-1)}-a_m\right)^2\to\min. \tag{9.2}$$

The coefficients $a_t^/$ are the values of a_t; the fixed values of these coefficients are determined based on the minimization of expression (9.2).

For configuring a single-layer neural network in m-chronology (x_n^m, y_n^m), one can select the nearest point k that maximizes the following expression:

$$p\left(x_t^m, x_n^m\right)+p\left(y_t^m, y_n^m\right),\ i=m, m+1,\dots n\to\max. \tag{9.3}$$

$$\begin{array}{c} x_{i1}^m, y_{t1}^m \\ x_{i2}^m, y_{t2}^m \\ \dots\dots\dots \\ x_{ik}^m, y_{tk}^m \end{array} \tag{9.4}$$

Using the classical method of least squares, we determine the forecast values for x_{n+1} and y_{n+1}:

$$\begin{aligned} x_{n+1}^f &= a_0^/x_n+a_1^/x_{n-1}+\cdots+a_{m-1}^/x_{n-(m-1)}+a_m^/, \\ y_{n+1}^f &= b_0^/y_n+b_1^/y_{n-1}+\cdots+b_{m-1}^/y_{n-(m-1)}+b_m^/. \end{aligned} \tag{9.5}$$

It is known that in time series x_t, there is a linear regression of x_{ir+1} to $\left|x_{ir}^m=\left(x_{ir}, x_{ir-1},\dots, x_{ir-(m-1)}\ (r=1,\dots k)\right)\right|$ [21]:

$$\sum_{r=1}^{k}\left(x_{ir+1}-a_0x_{ir-1}-\cdots-a_{m-1}x_{ir-(m-1)}-a_m\right)^2, \tag{9.6}$$

$$\sum_{r=1}^{k}\left(y_{ir+1}-b_0y_{ir-1}-\cdots-b_{m-1}y_{ir-(m-1)}-b_m\right)^2. \tag{9.7}$$

Depending on the nature of the forecasted object, the neural network can be single- or multi-layered. In our research, the method of a single-layer neural network is used, as illustrated in Fig. 9.1.

The single-layer structure of the neural network means that the neurons are connected in such a way that the input layer signal is transmitted directly to the output layer signal. That is, in this case, the neural signals of the input layer allow simultaneous necessary computations between the input and output layers and obtaining the desired result.

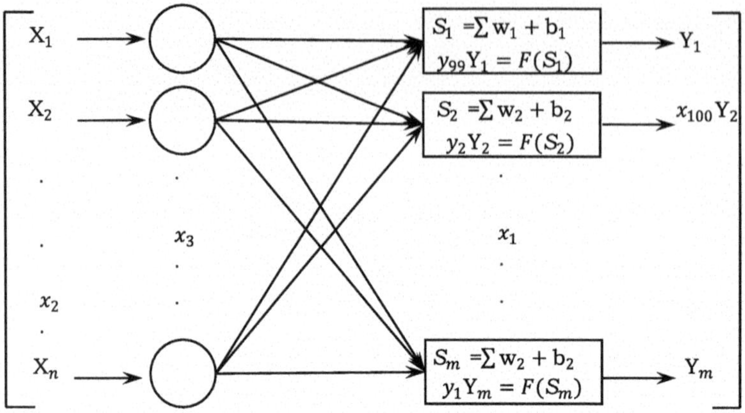

Fig. 9.1 Single-layer neural network

A single-layer neuron is capable of performing simple calculations, and the main function of the neural network is to perform computations as a result of the interaction of neurons from several input layers. In forecasting EPC using the artificial neuron method, a single-layer perceptron represents a simple network consisting of a group of neurons that make up this layer. Each new set of initial data entering the input layer is encoded using a vector of values and passed to the neuron of the next layer. In turn, neurons compute independently of each other [22].

It is known that SEPC is a relatively stable indicator, i.e., it does not change greatly over time. This means that the pattern of change in SEPC remains consistent regardless of the volume of production. This particular feature of such enterprises allows for determining with high accuracy the forecast indicators of SEPC. However, due to the nature of the technological process, some cases observe sharp changes in EPC volumes. For these cases, when choosing the composition and quantity of initial data for forecasting EPC, their in-depth analysis is required to avoid excessively abrupt changes in indicators. This, in turn, ensures a quality neural network training process.

It is known that in the training process, the output layer data are initially sorted. This process uses the sigma function. In the sorting process, non-stationary initial data are excluded, and the remaining data are moved to the next step. Data sorting is performed using the following expression:

$$f_t = \sigma\left(W_f\left[h_{t-1}, \alpha_t\right] + b_f\right), \tag{9.8}$$

where W_f—input weight coefficient, h_t—output vector; b_f—threshold of the input layer.

In the next process, the selected initial data are trained to determine the forecast indicator and their permissible values are identified. In this process, sigma and tangent functions are used:

$$i_t = \sigma\left(W_i\left[h_{t-1}, x_t\right] + b_i\right), \tag{9.9}$$

$$\tilde{C}_t = \tan\left(W_C\left[h_{t-1}, x_t\right] + b_C\right). \tag{9.10}$$

Here x_t—input layer of the time series, b_i and b_c—neuron thresholds.

Based on the selected data as a result of network research, the neurons in the output layer are identified through systematic combination of sigma and tangent functions. That is, in this process, the forecast result is determined:

$$\begin{cases} O_t = \sigma\left(W_O\left[h_{t-1}, x_t\right] + b_O\right) \\ h_t = O_t \cdot \tan(C_t) \end{cases}, \tag{9.11}$$

where O_t—original data selected using the sigma function;

h_t—output layer of the neuron.

When forecasting EPC using the ANN method, it is important to first choose the structure of the neural network, i.e., determine the number of input and output layers in the selected structure and the number of neurons in each layer.

Figure 9.2 shows the structure of a single-layer neural network for forecasting EPC at a metallurgical enterprise. The forecast is carried out using a single-layer neural network. In this case, as a parameter on the input layer, 100 protocol melts are introduced with data on EPC i.e., EPC is forecast based on the indicators of 100 output neurons corresponding to 100 input neurons.

Figure 9.3 presents a simplified block diagram of the neural network training algorithm in EPC forecasting. The training process is conducted using a supervised neural network with the backpropagation of errors algorithm. The neural network

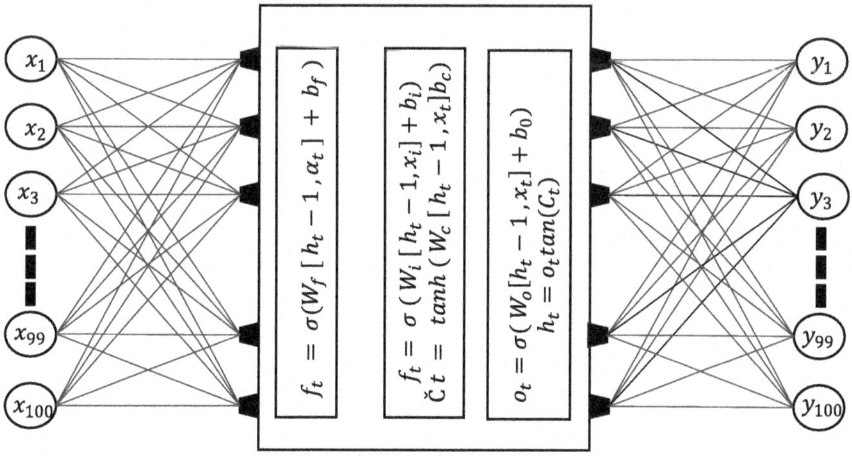

Fig. 9.2 Structure of a single-layer neural network for forecasting EPC

training process consists of two stages, in which the training is carried out depending on the composition and quantity of data initially present in the input layer.

In the first stage, the periodicity of incoming data entry into the database is annually. *In the second stage* of research, the periodicity of incoming data entry into

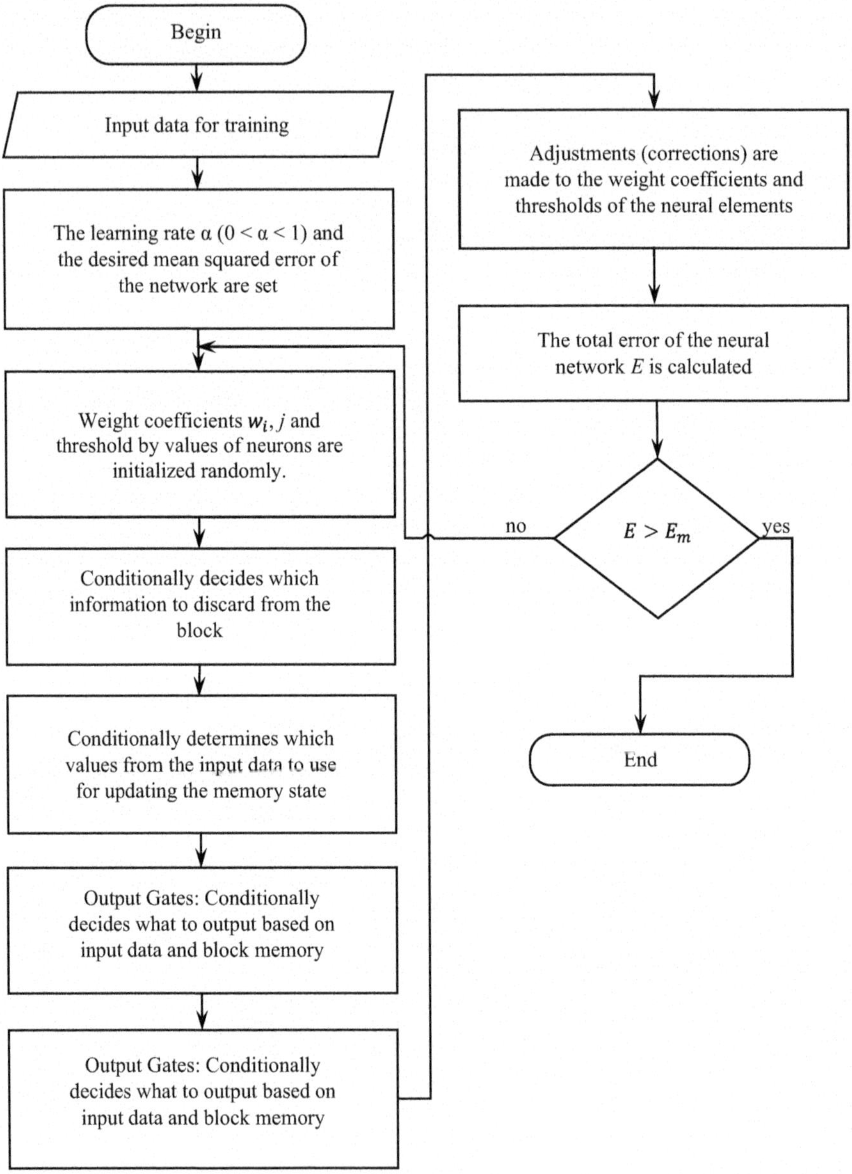

Fig. 9.3 Simplified block diagram of the ANN training algorithm for EPC forecasting

the database corresponds to days and months. At all stages, the duration of the learning process depends on the amount of data in the original database [23]. During the training of the neural network, it is tuned to reference values. This process is carried out by adjusting the weight coefficients, which in turn is based on minimizing functional errors.

The functional error, in turn, characterizes the difference between the reference value and the values of the output signal and is determined using the following expression:

$$E = \frac{1}{2}\sum_{p}\left(y_0^{(p)} - t_0^{(p)}\right)^2, \tag{9.12}$$

where p—the number of the input layer, $y_0^{(p)}$—the value of the output layer, $t_0^{(p)}$—the reference value.

The training process continues until the output values of signals from the neural network approximate the reference values. It concludes upon fulfilling the corresponding conditions of the algorithm, shown in Fig. 9.4.

Weight adjustment is one of the most important processes in training using ANNs; it is usually performed not for each input parameter but for their small groups. Such adjustment of weight coefficients accelerates the training process [24].

Although artificial intelligence provides highly accurate EPC forecasting results, it is very challenging to account for sudden spikes in EPC and similar natural phenomena. In this case, the forecast accuracy can be further enhanced by applying fuzzy logic to parts of the neural network responsible for data reception and printing the obtained results.

Fuzzy logic is a mathematical approach and a branch of logic that deals with analysis and decision-making in situations where there is ambiguity and uncertainty [25–28].

The dependency in Fuzzy logic is determined by two alternative methods:

1. If for each $f_j(t)$, m > 0 is an integer, where $f_j(t) = (f_{i1}(t-1) \times f_{i2}(t-2) \times \cdots \times f_{im}(t-m)) \circ R_a^p(t, t-m)$, and the Fuzzy dependency $R_a^p(t, t-m)$ is an indefinite relation, while $F(t)$ includes $F(t-1), F(t-2), \ldots, F(t-3)$, then the dependency is defined as follows:

$$F_j(t) = (F_{i1}(t-1) \times F_{i2}(t-2) \times \cdots \times F_{im}(t-m)) \circ R_a^p(t, t-m) \tag{9.13}$$

2. If $f_j(t) = (f_{i1}(t-1) \times f_{i2}(t-2) \times \cdots \times f_{im}(t-m)) \circ R_a^p(t, t-m)$ represents dependencies as $R_a^p(t, t-m)$, then F(t) is defined considering $F(t-1)$, or $F(t-2)$, …, or $F(t-m)$:

$$F_j(t) = (F_{i1}(t-1) \cup F_{i2}(t-2) \cup \cdots \cup F_{im}(t-m)) \circ R_a^p(t, t-m) \tag{9.14}$$

here, $R_a^p(t, t-m) = \cup_p R_0^p(t, t-m)$.

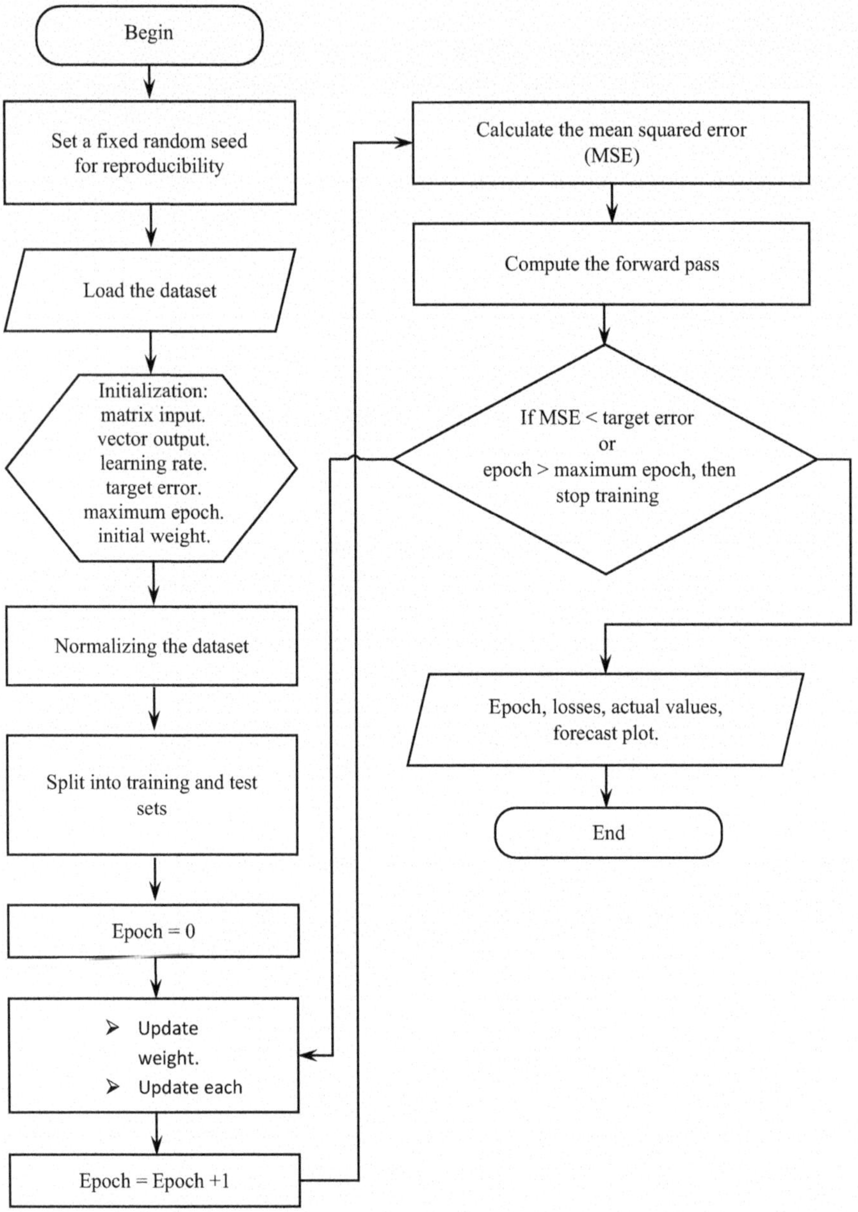

Fig. 9.4 Flowchart of the algorithm for constructing an EPC estimation model using the ANN method

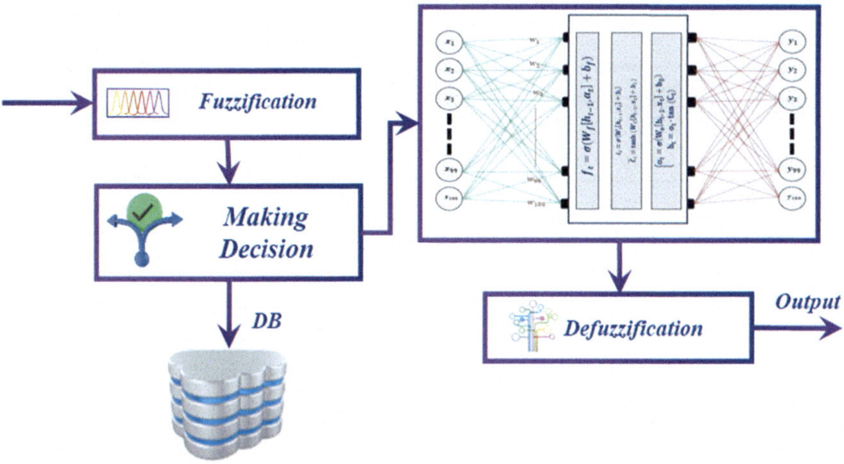

Fig. 9.5 Fuzzy 1 type mechanism

Fuzzy logic, in terms of operation, is divided into two mechanisms. In the Fuzzy 1 type mechanism, EPC forecasting is based on data acquisition and rule development based on a single data set (Fig. 9.5).

$$A = \int_{D_X} \mu_A(x)/x \tag{9.15}$$

In the Fuzzy 2 type mechanism, estimation is based on data acquisition and rule development using two or more types of data sets (Fig. 9.6).

$$\tilde{A} = \left\{ \left((x, u), \mu_{\tilde{A}}(x, u) \right) | \forall x \in X; \forall u \in J_x \subseteq [0, 1] \right\} \tag{9.16}$$

$$\begin{aligned} \tilde{A} &= \int_{x \in D_{\tilde{x}}} \int_{u \in J_x \subseteq [0,1]} \frac{\mu_{\tilde{A}}(x, u)}{(x, u)} = \int_{x \in D_{\tilde{x}}} \int_{u \in J_x \subseteq [0,1]} \frac{1}{(x, u)} \\ &= \int_{x \in D_{\tilde{x}}} \left[\int_{u \in J_x \subseteq [0,1]} \frac{1}{u} \right] 1/x \end{aligned} \tag{9.17}$$

$$\mu(x) = e^{-\frac{(x - x_0)^2}{2\sigma^2}} \tag{9.18}$$

Fuzzification. Fuzzification is the process of dividing the input data of a system into one or more linguistically logical fuzzy sets [29].

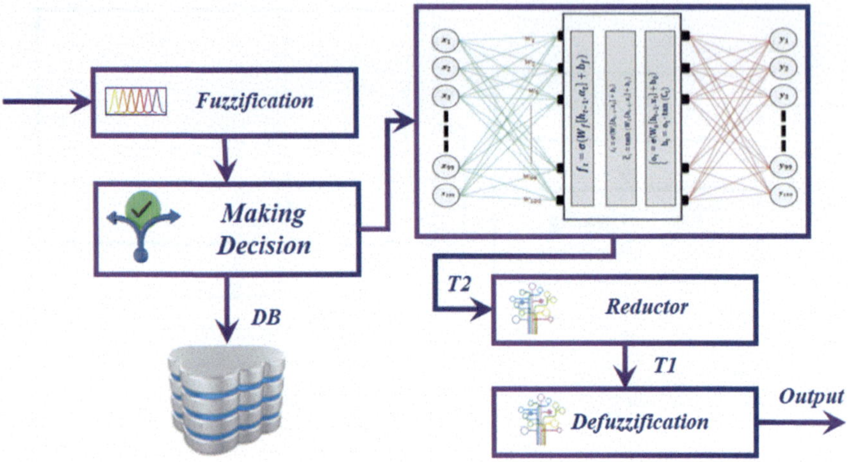

Fig. 9.6 Fuzzy 2 type mechanism

Fuzzy rule database. At this stage, expert knowledge is formulated in the form of a limited set of rules. The rule base contains those rules that will be used in decision-making. These rules are usually based on accumulated experience, but in some cases, rules can be derived using neural networks, genetic algorithms, or some empirical approach [30].

Decision making. In the process called decision-making, membership functions are combined with information from the fuzzy rule database to produce a final result.

Reducer. A component used to aggregate the output of several rules in the inference mechanism. In the fuzzy decision-making mechanism, several types of reducers are used in different places and with various functions:

1. Max–min reducer chooses the minimum output values from all rules.
2. Max reducer selects the maximum output values from all rules.
3. Sum reducer sends the sum of the output values from all rules to the next stage.
4. Weighted average reducer calculates the weighted average of all rule outcomes and sends the resulting average to the next stage.
5. Center of Gravity (COG) Reducer calculates the center of gravity of all rule outputs and weights according to the degree of reliability.

In the Fuzzy 2 type automatic analysis mechanism of EPC, the center of gravity reducer is applied.

Defuzzification. Defuzzification is the process of transforming fuzzy variables into precise values [29–32]. The defuzzification process takes fuzzy variables as input and prints a clear textual or numerical value as the output of the defuzzification process [33, 34].

1. In defuzzification, the following fuzzy sets are most commonly used and applied as output values.
2. Mean of Maximum Method: Determines the maximum value of the output membership function and uses the center of the maximum value as the output.
3. Bisector Method: The output identifies two points intersected by a line equal to half the maximum value of the membership function. The midpoint between these two is taken as the output value.
4. Weighted Average Method: Calculates the weighted average of the output membership function, using the output variable as the weight.
5. Height Method: Determines the maximum value of the output membership function and uses it as the output value.

Each defuzzification method has its advantages and disadvantages. The appropriate method is chosen based on the requirements of the task at hand.

Several defuzzification algorithms have also been developed, as listed below:

1. Mamdani Algorithm: One of the most widely used algorithms. When applying rules to input fuzzy sets, an output fuzzy set is generated, and the output membership function is determined using the maximum operator. Then, the output value is determined using methods such as centroid or mean of maximum (MOM).
2. Sugeno Algorithm: This method uses a linear function of input variables to determine output. The output membership function is a sum of weighted coefficients of input variables, and the weighted coefficients are determined by respective rules. The output is determined using the weighted average. This method also allows creating a mathematical model of data interrelation.
3. Tsukamoto Algorithm, which uses a continuous comparison function to compare input variables with output. The output membership function is determined by a combination of input variables and rules, and output data is determined by the centroid method.
4. Larsen Algorithm, similar to the Mamdani method, but it generates output membership functions using the logical multiplication ('product') operator instead of the maximum operator. Then the output undergoes defuzzification using the centroid method.
5. Hybrid Methods: In some cases, a combination of several algorithms is used to obtain results. For example, the Mamdani method can be used to create output fuzzy sets, and the Sugeno method for determining the output membership function.

Figure 9.4 shows the flowchart of the algorithm for constructing estimation models of EPC using the ANN method, detailing the stages of model development using Python 9.9, the most relevant and demanded programming language currently.

As seen in the flowchart (Fig. 9.3), the main condition for the success of forecasting using the ANN method is the correct choice of Epoch indicators and mean squared error. This flowchart is based on a universal algorithm for constructing a model for forecasting EPC using the ANN method. It can be used not only for forecasting EPC of the studied object but also by other industrial enterprises.

Thus, the use of artificial neural networks in forecasting EPC allows for calculating forecast indicators with high accuracy. This is primarily because this method has the ability to learn and adapt to the process of forecasting EPC. Developing forecasting models using this method is a more complex process compared to the methods discussed above. The correct selection of the composition of input layers and exclusion of various potential cases ensures the successful completion of calculations. This is also evident from the analysis of the error graphs of indicators determined using this methodology.

References

1. Johnson LE, Patel SR (2022) Estimation method for industrial enterprise electrical power consumption. J South-Ural State Univ Ser Comput Technol Manage Radioelectron 16(1):6–14
2. Green KR (2021) Global and Uzbekistan's electric power industry: a comparative analysis. Science and Technology, Tashkent, p 470
3. Miller AA (2020) Multilevel planning and management methods for electrical power consumption at a metallurgical enterprise, p 130
4. Smith VF, Brown VA (2022) Mathematical modeling of electrical power consumption by production systems. Electr J 3:22–28
5. Thompson NN (2021) Methodological aspects of energy saving and enhancing energy efficiency in industrial enterprises: a monograph. Udmurt University Publishing, Izhevsk, p 120
6. Bozdogan H (2022) Statistical data mining and knowledge discovery: new applications in various fields. Chapman & Hall/CRC, p 600
7. Williams IV (2022) Estimation industrial electrical power consumption with artificial neural networks. Kemerovo, p 150
8. Martin AV (2023) Analysis of current and future methods for estimation energy resource consumption at industrial enterprises. Ind Power Eng 3:18–21
9. Amosov AA, Dubinsky YA, Kopchenova NV (2022) Computational methods for engineers. Moscow Power Engineering Institute, Moscow, p 600
10. Zarubin VS (2023) Mathematical modeling in engineering. Bauman MSTU Publishing, Moscow, p 500
11. Solovyeva IA, Dzyuba AP (2023) Estimation electrical power consumption considering technological and market environment factors. Sci Dialog 7(29):95–110
12. Kazarinov LS, Barbasova TA, Zakharova AA (2023) Predictive control method for energy efficiency of an industrial enterprise. J South-Ural State Univ Ser Comput Technol Manage Radioelectron 2:10–22
13. Filimonova AA (2023) Multilevel planning and management of electrical power consumption at a metallurgical enterprise. In: Proceedings of the all-Russian forum information society-2023: challenges and tasks. Publishing Center of SUSU, Chelyabinsk, pp 88–97
14. Kazarinov LS, Barbasova TA, Zakharova AA (2023) Optimal estimation of energy resource consumption based on cost criteria. J South-Ural State Univ 15(1):88–92
15. Barikaev EN, Chernyak VZ (2023) Methods of expert evaluations. Bull Mosc Univ Ministry Intern Aff Russ 12:182–188
16. Bolshakov AA, Karimov RN (2002) Methods for processing multidimensional data and time series: a university textbook. Moscow, p 520
17. Kalinina VN, Pankin VF (2002) Mathematical statistics. Tutorial, Moscow, p 340 (in Russian)
18. Rakhmonov IU, Niyozov NN (2019) Analysis of existing methods of electric consumption. Sci J Eur Sci Rev Austria Vienna (9–10):209–211

19. Madala HR, Ivakhnenko AG (1994) Inductive learning algorithms for complex system modeling, p 368
20. https://controlengrussia.com/perspektiva/neural_networks_/
21. https://www.stackvidhya.com/train-test-split-using-sklearn-in-python/
22. https://algotrading101.com/learn/train-test-split/
23. Fang W, Jiang J, Lu S, Gong Y, Tao Y, Tang Y, Yan P, Luo H, Liu J (2020) A LSTM algorithm estimating pseudo measurements for aiding INS during GNSS signal outages. Remote Sens 256. https://doi.org/10.3390/rs12020256
24. https://www.upgrad.com/blog/types-of-supervised-learning/#5_Neural_Networks
25. Kayacan E, Khanesar MA (2016) Fuzzy neural networks for real-time control applications, 2nd edn, p 447
26. Hossain MA, Shill P, Sarker B, Murase K (2012) Optimal fuzzy model construction with statistical information using genetic algorithm. Int J Comput Sci Inf Technol 3(6):241–247. https://doi.org/10.5121/ijcsit.2011.3619
27. Kalnin N (2017) Hybrid fuzzy logic and extremum seeking attitude control of solar sail spacecraft, p 166
28. https://www.energycap.com/
29. Rakhmonov IU, Najimova AM, Esemuratova ShM, Koptileuov TT (2022) Development of a method for determining the main and additional factors affecting the forecast of electrical power consumption. AIP Conf Proc 2649:070024. https://doi.org/10.1063/5.0104793
30. Rakhmonov IU, Niyozov NN, Kurbonov NN, Umarov BS (2023) Estimation of electrical power consumption by industrial enterprises with a continuous nature of production. E3S Web Conf 384:01030-1–01030-4. https://doi.org/10.1051/e3sconf/202338401030
31. Rakhmonov IU, Najimova AM (2022) Forecast of electrical consumption in cotton-textile cluster enterprises taking into account the volume of production. AIP Conf Proc 2647:030011. https://doi.org/10.1063/5.0104791
32. Rakhmonov IU, Najimova AM, Reymov KM (2022) Cotton-textile cluster enterprises use the small square method in estimation electrical power consumption. AIP Conf Proc 2648:030010. https://doi.org/10.1063/5.0104788
33. Rakhmonov IU, Hoshimov FA (2020) Development of an algorithm for evaluating the dominant factors that have the greatest impact on the energy intensity of products. E3S Web Conf 209:07018. In: Sustainable development and smart management, ENERGY-21. https://doi.org/10.1051/e3sconf/202020907018
34. Rakhmonov IU, Tovbaev AN, Nematov LA, Alibekova TSh (2020) Development of forecasted values of specific norms for the issues of produced products in industrial enterprises. J Phys Conf Ser 1515:022050. In: ICMSIT-2020. https://doi.org/10.1088/1742-6596/1515/2/022050

Chapter 10
Reserves of Energy Saving and Improvement of the Energy Efficiency

This chapter deals with the potential for energy saving and improvement of the energy efficiency (ES and IEE), methods for its assessment, and the economic efficiency of measures to realize it. ES and IEE measures are classified. A methodology is proposed for assessing the impact of technological and organizational factors on energy production indicators, which allows their justified inclusion in the reserve for saving energy resources.

10.1 Estimation of the Economic Efficiency of Energy Saving Measures

As the power capacities of the fuel and energy complex (FEC) in developed and developing countries increase due to the growing energy consumption in production and domestic sectors, the importance of ES and IEE becomes increasingly significant. Rational use of electricity by the most energy-intensive consumers—industrial enterprises—not only improves their economic indicators but also contributes to increasing the volume, quality, and cost-effectiveness of their produced goods.

Identification of the fuel and energy saving reserves in enterprises and production complexes should focus on reducing losses in electrical networks, rationalizing technological processes, introducing new equipment, modernizing existing equipment, more fully utilizing low-potential secondary heat, and improving the rationing and control of rational electricity and heat energy consumption [1].

Reducing energy consumption in enterprises does not always require complex technical measures; in many cases, simply adhering to the standard regime of electricity usage is sufficient, involving timely turning off lights, preventing idle operation of equipment and units, etc. This indicates that fuel and energy resources (FER) saving reserves exist in every enterprise.

I. U. Rakhmonov et al., *Electric Consumption by Industrial Enterprises*, Power Systems, https://doi.org/10.1007/978-3-031-62676-0_10

Energy saving is not an end in itself, and saving FER should not be pursued at any cost. In the worst-case scenario, financial investments in the development and implementation of energy-saving programs/measures might exceed the economic benefits from energy saving. To prevent financial losses in forming a set of energy-saving measures, it is essential to develop methods for assessing the effectiveness of ES and IEE programs with optimized initial investments intended for their implementation. In turn, forming optimized ES and IEE programs requires classification of energy-saving measures. Currently, two main interrelated classifications are proposed: (a) by duration of implementation, payback periods, and economic effect, (b) by the nature of measures: organizational, technical, technological, etc.

Figure 10.1 presents information on the implementation of a large number of energy-saving measures, the details of which are borrowed from available sources and summarized in [2].

Group I—Highly efficient short-term measures (less than 1 year), which include organizational (low-cost) measures:

- development and maintenance of technical and energy passports;
- rationing (standardization) of EPC for different types of activities, including maintaining acceptable sanitary and hygienic conditions for service personnel in technological premises;
- increasing the level of awareness among staff and the public in the field of energy saving;
- granting consumers the right to manage funds saved through energy saving, primarily for staff incentives and creating an investment fund for measures requiring more substantial funding;

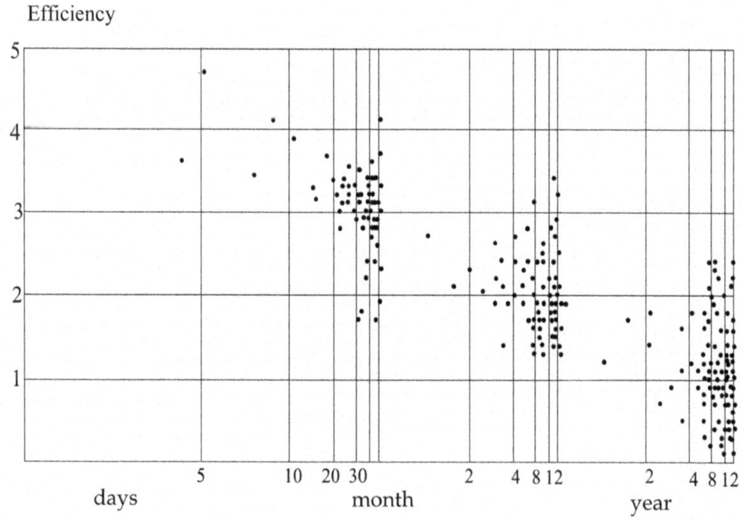

Fig. 10.1 Dependency of the economic efficiency of energy saving measures on their implementation period

- establishing procurement rules that encourage the use of energy and efficient technologies;
- reorganization of municipal heating supply companies into commercial enterprises or public and private partnerships.

This group may also include some organizational, technical, and rationalizing energy saving measures with a payback period of 5–6 months:

- adherence to energy efficient technological modes and work schedules;
- replacing energy equipment of excessive capacity with equipment of nominal capacity;
- equipping consumers with resource meters, including air, effluent, discharges, etc.

The share of low-cost measures can account for 30–50% of the total energy saving potential of an enterprise.

Group II—Medium-efficiency and medium-term measures, which include both substantial organizational measures and technical/technological measures:

- development and implementation of energy efficiency standards for the operation of buildings and industrial equipment, fuel use;
- coordinated plans for heat supply;
- stimulating banks and leasing companies to finance energy efficient projects;
- conditioning the granting of subsidies for major repairs on commitments to increase energy efficiency;
- use of reactive power compensation units;
- reducing the power of heaters due to insulation and repair of premises;
- tuning and optimizing the operation of units;
- replacement of morally and physically obsolete equipment with modern energy efficient equipment.

This group of measures requires significant investment and is implemented in cases where the effect of ES and IEE within acceptable periods (no more than 2–3 years) will cover the costs of their implementation.

Group III—Long-term, high-cost organizational and technical/technological measures, which envisage:

- Reform of tariff formation.
- Liberalization of electricity and gas markets.
- Development and implementation of a comprehensive set of measures aimed at enhancing the energy and ecological efficiency of energy resources and carriers. These measures include: radical reconstruction of production, major renovation of residential and industrial buildings to improve the thermal resistance of enclosing structures and modernization of external and internal engineering communications; optimization of the fuel and energy balance (FEB) of consumers, including through the expanded use of unconventional renewable energy resources.

To implement these measures, external borrowed funds are typically required. The payback period for these activities often turns out to be long, and therefore their economic efficiency is relatively low.

Modern methods for forming and determining the sequence of implementing a set of energy-saving measures take into account their ranking and temporal optimization, and are based on the fundamental principles of financial and economic analysis and the assessment of the effectiveness of investments.

The level of economic losses at the current time due to the postponement of energy-saving measures can be estimated by an index (K_{los}), reflecting the ratio of the reduction in net discounted revenues in the subsequent period (ΔD) to the volume of investments in energy saving (Es):

$$\kappa_n = \frac{\Delta D}{E_s}. \tag{10.1}$$

This method minimizes the subjective factor inherent in the widely used expert evaluation method. In this case, experts are involved to determine the reliability of the performance indicators of the measures.

Economic losses caused by the lack of a consistent implementation of an energy saving policy on a national, regional, municipal level, etc., also require economic evaluation. This can be conducted using a ranking criterion considering all revenue and expenditure items (reducing payments for environmental pollution, budget replenishment, etc.). Losses assessed based on this indicator from possible delays of not just one, but the entire set of available energy saving measures, provide a more accurate estimate of the energy saving potential (ESP) of the chosen object.

In rigorous calculations, a parameter—*the technical potential of energy saving*—is used. This potential is defined by the achievable current levels of energy efficiency of equipment and technologies, limited by physical principles.

In technical and economic calculations, the technical potential for energy saving is assessed by comparing, for each economic sector, the efficiency level of technologies used in a particular country with the efficiency level of market-available technologies used in leading countries.

The technical potential may be unsuitable for implementation as an investment project if investments in ES and IEE are too large or provide too little savings over the lifetime of the investment project to be attractive to energy consumers. Investments must be *financially* attractive and economically expedient.

The estimation of the economic viability of investment projects in ES and IEE is based on calculating efficiency criteria, the main ones being:

- net present value,
- return on investment,
- internal rate of return,
- payback period of the investments.

Overall, it is similar to the evaluations of any investment projects.

Financially attractive investments are capital expenditures that lead to savings of energy resources and funds for end consumers—investors, due to the fact that the cost of saving a unit of energy (for example, 1 kWh) is less than the cost of acquiring an additional unit of energy. The cost of saving a unit of energy depends on initial capital expenditures, potential additional costs for operation and maintenance to achieve energy savings, and the investor's alternative opportunities, besides improving energy efficiency. The cost of acquiring an additional unit of energy is determined by the tariff or market price of the energy resource (depending on whether the prices of energy resources are regulated or not) for these investors.

For private investors, particularly households (individual investors), higher discount rates are applied. This is due to three reasons: (a) households are generally more reluctant to engage in the risk associated with investments in energy efficiency, (b) for investing, they often have to take out loans at interest rates set for private borrowers, (c) from the perspective of household owners; there are many other areas for investing their disposable funds.

Economically appropriate investments are those capital expenditures that lead to savings in energy resources and overall financial resources over the lifetime of the investment project. However, the savings or revenue from the sale of freed-up resources may not necessarily be realized by any specific energy consumer or provide an attractive level of income for a private investor. Such investments are prudent for the government in the interest of the country but may not be appropriate for individual investors. Investments are considered economically appropriate if the cost of saving a unit of energy (for example, 1 kWh) is less than the costs incurred by the state for constructing a unit of new generating capacity (for example, 1 kW) or than the opportunity cost for the country from exporting a unit of energy resource, depending on which of these values is greater.

Economically efficient investments differ from financially effective ones in that they take into account external effects, both positive and negative. The most significant external effect is the reduction of carbon dioxide (CO_2) emissions. According to the mechanisms of the Kyoto Protocol implementation, energy savings achieved in certain economic sectors can be sold in the form of CO_2 emission quotas.

In estimating the economically appropriate potential of energy saving, other positive external effects, such as the reduction of emissions of pollutants other than greenhouse gases, should also be considered. Considering these effects, along with rising energy resource prices and decreasing capital costs, increases the share of economically appropriate investments.

10.2 Methods for Evaluating the Effectiveness of Electricity Saving Measures

A significant part of ES and IEE measures is inherently linked with the rationalization of existing and the introduction of new technology, improvement of production organization, increase in the volume of produced goods, and enhancement of the quality indicators of their production.

It is evident that in developing and implementing each ES and IEE measure, the evaluation and justification of their economic feasibility and effectiveness should be conducted only in conjunction with the entire complex of technological, energy, and organizational-technical factors. Incorrect assessment of the reduction in electrical energy costs because of the implementation of a particular measure can lead to a distortion in the total cost and, consequently, to erroneous results.

For determining the feasibility of ES and IEE measures, enhancing the power utilization factor, and reducing electricity losses, outdated methods are often used. Clearly, an assessment of measures based solely on the amount of electricity saved, without considering such important factors as changes in operating conditions and capital expenditures associated with the implementation of the measure is flawed. Furthermore, the absolute amount of electricity savings cannot be a comprehensive indicator of the effectiveness of the measure if considered in isolation from the operating mode of the equipment on which the measure is implemented, either at maximum load or at a forced reduction due to limitations from the energy supply organization.

All electricity saving measures at industrial enterprises can be divided into the following groups [3–5]:

1. Measures to reduce power consumption.
2. Measures to reduce energy consumption without decreasing productivity.
3. Measures to increase the consumption of electric power and enhance productivity.
4. Measures to increase productivity and energy consumption.
5. Measures to identify productivity without changing power consumption and energy.

These measures can be divided into two groups:

(a) Measures not associated with the volume of produced goods;
(b) Measures associated with the volume of produced goods.

In calculations related to increasing the power factor and reducing active losses in electrical machines and devices, the aim is to achieve a minimum of allocated losses, i.e., the minimization of total active power losses in the electrical installations themselves and losses of active power in the power system networks caused by the reactive load of these machines and devices.

Allocated losses cannot be the basis for evaluating measures in the context of a self-accounting enterprise, as it is known that in the technical–economic indicators of an enterprise, only the energy directly consumed in the enterprise's energy management can be reflected.

Electricity savings in the context of an industrial enterprise are achieved by reducing electricity losses in networks, devices, machines, and by lowering the SEPC.

The annual electricity savings on a given technological unit, process, device, etc., can be determined from the expression:

$$W_{es} = (P_a - P_n) \cdot \tau = \Delta P_{es} \cdot \tau, \text{ kWh/year} \tag{10.2}$$

or

$$W_{es} = (d_a - d_n) \cdot V_n, \tag{10.3}$$

where P_a, P_n are the average consumed power, respectively, before and after the implementation of the measure;

d_a, d_n—the SEPC per unit of product, respectively, before and after the implementation of the measure;

ΔP_{es}—productivity of the units, machines, processes, etc., one year after the implementation of the measure;

τ—number of working hours in a year.

The calculated amount of electricity savings (plant-wide) that should be considered in the calculation of the cost of factory products must include the reduction of electrical energy losses in transformers, converters, in workshop and factory networks up to the point of separation of the networks from the power system. The magnitude of this savings is:

$$W_{p.es} = W_{es}\frac{1}{\eta_p}, \ \Delta P_{p.es} = \Delta P_{es}\frac{1}{\eta_p}, \tag{10.4}$$

where η_p—the resultant efficiency of transmission and transformation of electrical energy in the factory network from the electrical receiver to the point of separation of the factory network from the power system network.

Since costs for electricity, fuel, and raw materials are included in the calculation at the prices at which they are released to the enterprise, naturally, the effect of implementing measures, which is reflected in the cost of production, cannot be evaluated differently than by the tariff.

10.3 Methodology for Estimation the Potential for Electricity Savings

Experience from research and experimental work conducted in electric power and electrical power technology processes on existing equipment of enterprises in various industrial sectors using system analysis methods demonstrates the high efficiency of these works and the scientific validity of the recommendations proposed in [6, 7].

The main directions of analysis and estimation of the reserve for reducing EPC by industrial enterprises are:

- identifying patterns in the change of electrical loads of electric drives with changes in the operating mode of technological equipment;
- developing corresponding mathematical models that reveal possibilities for minimizing electrical loads;
- estimating the impact of energy, technological, and operational factors on the magnitude of electrical loads, absolute and SEPC;
- using equations of pairwise and multiple correlations, as well as graph-analytical methods;
- identifying functional dependencies of electrical parameters on the design features of electrical and thermal units;
- developing methods for the most advantageous distribution of power in the high-temperature zone inside furnaces;
- developing methods to reduce electricity losses in industrial electrical networks, electrical machines, and devices;
- rationalization of reactive power generation and distribution, considering voltage levels in accordance with static characteristics;
- development of new methods for regulating electrical loads with reduced EPC during peak load hours of the power system, as well as for enterprises operating around the clock without reducing their productivity;
- exploring ways and methods to convert enterprises into consumers-regulators of power system loads;
- development of algorithms and a set of programs for analysis, calculation, forecasting, and rationing of energy resources, taking into account both internal and external inter-industrial connections;
- analysis of energy resource savings when switching to energy saving modes of existing equipment and implementing new, prospective machines, mechanisms, and technological processes.

10.4 Examples of Electricity Savings at Industrial Enterprises in Various Sectors

Constructing energy characteristics allows for monitoring the dynamics of loads and SEPC of each unit and shop of enterprises. In this context, energy characteristics are dependencies of active power P, total W, and SEPC e on the volume of primary raw materials G or produced goods PG [8]. Energy characteristics enable the establishment of rational modes corresponding to minimal values of expenditures and norms of electrical energy and determine the potential for reducing EPC.

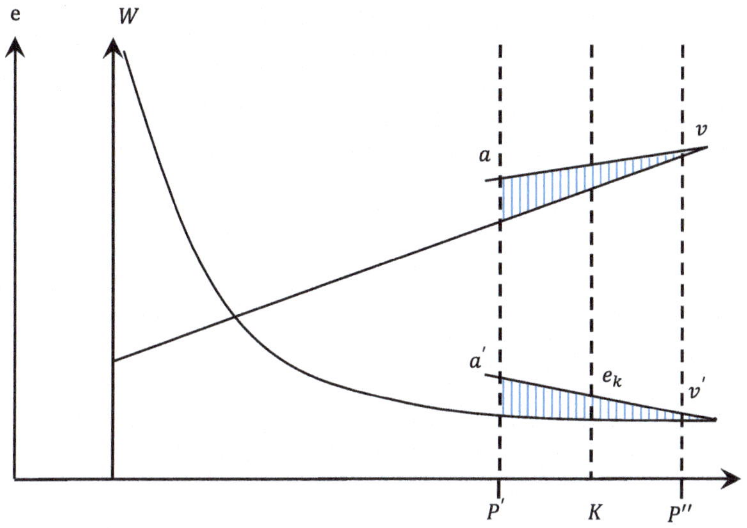

Fig. 10.2 Normative energy characteristics of an enterprise

For a workshop and the enterprise as a whole, energy characteristics are constructed based on primary experimental or calculated data. Based on experimental data for enterprises and individual workshops and technological lines, real consumed powers and volumes of electrical energy are determined, which serve as primary data for constructing energy characteristics [7].

In Fig. 10.2, the curves $W = f(P)$ and $e = f(P)$ represent the minimum possible level of EPC and SEPC, which cannot be further reduced with the current composition and technical level of the technological equipment. Here, each productivity point corresponds to a strictly defined number of machines, mechanisms, processing machines, lines, etc., operating at maximum productivity.

When determining the fixed part of EPC, expenses for lighting load, heating, ventilation, and other factors are considered. Using methods of mathematical statistics and real industrial production data, the limits of EPC corresponding to a certain segment of the energy characteristic are determined. These are in turn confined within the possible range of changes in the quantity of produced goods, reflecting the level of EPC $(a - v)$, $(a' - v')$.

Consequently, the values of the potential for reducing EPC for point K can be calculated using the formula:

$$\Delta W_{es} = (e_k - e) \qquad (10.5)$$

The potential for reducing electricity consumption, determined by formula (10.5), is general for a certain operation and organization of production. To determine the actual value of the EPC, it is necessary to identify the factors that cause unproductive EPC. In this case, correlational and multi-factor regression analyses can be applied.

Significant factors influencing the indicators of EPC are determined based on the analysis of information obtained with the help of enterprise specialists.

Here, the primary criterion is the parameter of SEPC.

Considering the above, the mathematical expression for special electrical power consumption is:

$$e = a + bx_1 + cx_2 + \cdots + \gamma x_n, \tag{10.6}$$

where $x_1, x_2, \ldots, x_n$—are significant normative factors: equipment utilization coefficient, temperature, humidity, etc.; a, b, c, γ—constant coefficients.

The developed model is used to identify those factors that not only qualitatively assess but also accurately determine the potential for reducing EPC.

Using the characteristics $e = f(P)$, another task such as identifying the reserve for reducing energy costs at enterprises where the useful product is extracted from raw materials and the quality of the finished product depends on its content in the raw material and the degree of extraction (phosphorites, raw cotton, etc.) can also be addressed. Naturally, increasing the yield coefficient (ξ) of the useful product and reducing its losses can provide significant savings in consumed energy [8].

The EPC per unit of processed raw material is:

$$e_G = \frac{W_G}{G}, \tag{10.7}$$

where W_G—is the EPC per unit of processed raw material.

G—is the amount of raw material processed.

The quantity of finished product manufactured is:

$$V = \xi G \tag{10.8}$$

Correspondingly, the SEPC for the production of the finished product is found from the expression:

$$e = \frac{W}{\xi G} \tag{10.9}$$

Assuming that electricity consumption remains constant when processing the same volume of raw material and does not depend on the content of the useful product and the yield coefficient ($W_G = W$) we obtain:

$$e = \frac{1}{\xi} e_G \tag{10.10}$$

The value (e) can also be obtained directly from the energy characteristic $e = f(P)$.

An increase in the yield coefficient due to the improvement of technology and organization of production by the amount $\Delta\xi$ will lead to a decrease in SEPC and electricity savings in accordance with the expression:

$$\Delta e = \left(\frac{1}{\xi} - \frac{1}{\xi + \Delta\xi} \right) e_G \tag{10.11}$$

For enterprises where the process of extracting the useful product from raw material is two-stage (preliminary and complete extraction), the reserve for energy savings lies in the ratio of the yield coefficients of the useful product at each stage. For example, in oil extraction from seeds in a pressing production, part of the oil remains in the shell and is extracted by extraction in the second stage of the process. The SEPC at each of the two stages will change oppositely depending on the change in the extraction coefficient at the first stage—the higher it is in the preliminary pressing, the lower the SEPC and vice versa—higher in the extraction process. The energetically most favorable coefficient within the technological regulations can be obtained by a graph-analytical method, taking into account the indicators of both stages of the oil extraction process [7].

Significant reserves for saving energy resources are hidden in measures to reduce the material intensity of the finished product, as well as reducing the special indicators of raw material and semi-finished product consumption.

To solve this problem, it is necessary, firstly, to have a clear understanding of the energy intensity of the materials and products used, and manufactured products, which is not always possible. The fact is that to date, there is no unified approach to assessing the energy intensity of individual types of industrial products. For example, for enterprises whose complex technological process includes a cycle from processing primary raw materials to manufacturing finished products, it is necessary to account for and determine the total SEPC for the volume of product manufactured.

For enterprises with a discrete production cycle that obtain semi-finished products from external enterprises, the electrical intensity of the finished product is determined only by the energy consumed by this enterprise. Consequently, the referenced parameter of SEPC for identical final products varies widely. Furthermore, the method of assessing energy intensity within the enterprise itself also varies. For example, if we consider enterprises of different industries with a complete production cycle, such as those producing cement and cottonseed oil, in the first case, the energy intensity of the finished product (cement) includes the production of the semi-finished product—clinker, while in the second case, the energy intensity of the finished product (refined oil) does not include the energy intensity of the semi-finished product (crude oil). As a result of the lack of a unified approach to solving this issue, in some cases, there are significant difficulties in assessing the true energy intensity of a given product.

Let is consider the method of evaluating the complete energy intensity of an industrial product, in the manufacturing of which several enterprises, including those from other industrial sectors, participate.

Let the total number of productions involved in obtaining and manufacturing the raw materials, semi-finished products, and materials necessary for the production of the final product be denoted by K; the volume of the final product by Z; the volume of products manufactured by each of these productions by $Q, Q_1, \ldots, Q_n$.

The complete electrical intensity is given by:

$$\varepsilon_n = \sum_{1}^{m} \sum_{1}^{n} \varepsilon_i \gamma_z + e, \tag{10.12}$$

where m—the number of types of raw materials, materials, semi-finished products entering the enterprise for the manufacture of the final product; ε_i—the SEPC for the production of the ith production involved in the release of the final product; $\gamma_{zi} = \frac{Q}{Z}$—the special consumption of the ith intermediate product for the production of a unit of final product; e—the value of SEPC at the enterprise manufacturing the final product.

Reducing the consumption of raw materials, materials, and semi-finished products in the manufacture of finished products is a very important direction in the work on the rational use of energy resources. These costs are typically not considered in electric power calculations. However, they are decisive in the volume of EPC at all the numerous technological stages of each production. It should be noted that the quality of raw materials and materials, the technical condition of the technological equipment determine the SEPC. Their consideration will elevate the level of work on rationing SEPC to a new qualitative level.

In the developed method, the specific consumption of raw materials and materials are components of the calculated expression for SEPC for each piece of technological equipment, workshop, and the enterprise as a whole. As a result, potentials for reducing SEPC are determined through the optimization of parameters of raw materials, and energy.

For example, consider data from a chemical enterprise.

The electricity consumption per ton of ammophos (factory indicator) is, for:

ammonia $\gamma_a = 0.322$;
phosphorites $\gamma_{ph} = 0.322$;
sulfuric acid $\gamma_{sa} = 0.322$.

Accordingly, the SEPC at enterprises producing these materials is:

$\xi_a = 13{,}640$ kWh/t (electrolytic ammonia);
$\xi_{ph} = 50$ kWh/t;
$\xi_{sa} = 87$ kWh/t.

The total special electric intensity of ammophos is:

$$\xi_{sa} = 13{,}640 \cdot 0.327 + 50 \cdot 0.425 + 87 \cdot 0.322 + 620 = 5060\ \text{kWh/t}$$

With $\xi_a = 1580$ kWh/t using conserved ammonia in technology

$$\xi_a^n = 1197\ \text{kWh/t}$$

Thus, the complete (true) electric intensity of ammophos is 2–8 times greater than the electric intensity calculated by the factory—the manufacturer of the final product.

The amount of electricity savings from saving the final product by Δz in its use is:

$$W_{es} = \xi \cdot \Delta Z \quad (10.13)$$

In our example, each ton of saved ammophos provides more than 5000 kWh of electricity savings.

Next, we consider the method of assessing the impact of material intensity on the electric intensity of the finished product and the amount of energy-saving potential from reducing this indicator.

The overall electricity savings at the enterprise from reducing the material intensity in workshops is:

$$W_{es} = \sum \left[\alpha_i - (d_i \pm \Delta d_i)(q_i - \Delta q_i)\right] Z \quad (10.14)$$

Here q and Δq are the specific consumption of raw materials or semi-finished products and, respectively, deviation per unit of final product,

d and Δd—the SEPC and, respectively, its deviation per unit of workshop product,

$a = dq$ is the share of this workshop in the amount of SEPC consumption in the production of the final product.

Thus, the proposed methods of electric power estimation of the influence of technological and organizational factors on the energy indicators of production allow for a reasoned inclusion in the reserve for saving energy resources and contribute to the improvement of existing and the creation of new energy-saving processes and types of equipment. Furthermore, the magnitude of the total electric intensity of the final product of industrial enterprises and the potential for its reduction should be considered at the stages of product manufacturing, planning, and forecasting energy indicators.

References

1. Ushakov VYa, Kharlov NN, Chubik PS (2015) Problems of energy saving and its implementation at fuel and energy enterprises. Tomsk Polytechnic University Publishing House, Tomsk, p 283 (in Russian)
2. Ushakov VYa. Energy saving and increasing energy efficiency: socio-economic, organizational and legal aspects: textbook. Tomsk Polytechnic University Publishing House, Tomsk, p 280 (in Russian)
3. Hasanbeigi A (2010) Energy-efficiency improvement opportunities for the textile industry. In: International energy analysis, 2010. Retrieved from https://international.lbl.gov
4. Deng Y, Cao G, Wei Z, Xu X, Ma L, Gao G (2022) Investment merger efficiency evaluation of energy enterprises. From a perspective of energy internet. Front Environ Sci 10. https://doi.org/10.3389/fenvs.2022.1036772
5. Rakhmonov IU, Ushakov VYa, Niyozov NN, Kurbonov NN, Mamutov M (2021) Energy saving in industry. E3S Web Conf 289:07014. In: Energy systems research 2021. https://doi.org/10.1051/e3sconf/202128907014

6. Rakhmonov IU, Ushakov VYa, Najimova AM, Jalilova DA, Omonov FB (2021) Minimization of energy production management consumptions. E3S Web Conf 289:07013. In: Energy systems research 2021. https://doi.org/10.1051/e3sconf/202128907013
7. Rakhmonov IU, Kurbonov NN (2020) Analysis of automated software for monitoring energy consumption and efficiency of industrial enterprises. E3S Web Conf 216:165717. In: Rudenko international conference on methodological problems in reliability study of large energy systems. https://doi.org/10.1051/e3sconf/202021601178
8. Hoshimov FA, Rakhmonov IU, Niyozov NN (2020) Technology to reduce energy costs in the electric steel melting shop. E3S Web Conf 209:07017. In: Sustainable development and smart management, ENERGY-21. https://doi.org/10.1051/e3sconf/202020907017

Chapter 11
Ways to Reduce Electric Energy Consumption

This chapter focuses on the issues of electrical power saving resulting from the reduction of raw materials and semi-finished product consumption, selecting criteria for the optimality of energy indicators in energy-intensive units and units of mass use. The main directions for reducing specific electrical power consumption (SEPC) acceptable for enterprises in practically all industrial sectors are analyzed.

11.1 Electrical Power Saving Due to Reduction in Raw Material and Semi-finished Product Costs

Energy saving measures are effective only when technological and operational factors affecting the energy intensity of industrial products are considered. For this purpose, it is necessary to create a unified information base for specific sectors of the national economy, characterizing technical, technological, organizational, and economic aspects of energy resource transformation, distribution, and use. Reducing the expenditure of raw materials, materials, and semi-finished products for the production of finished goods is an important direction in the work on the rational use of energy resources. Currently, the costs of electric energy resources are, as a rule, not accounted for in the numerous technological processes of each production [1].

The level of SEPC depends on the quality of raw materials and materials, the technical perfection of technological equipment and processes. To produce 1 ton of finished product, it is sometimes necessary to process dozens of tons of raw materials and materials, and where the organization of production is not perfect, there can be excessive losses and, consequently, an overuse of energy. Considering all this, the improvement of EPC norms receives significant attention in this monograph.

The essence of the method proposed by the authors lies in the fact that the specific costs of raw materials and materials are directly included in the calculation formulas for SEPC of each production unit of the enterprise. In this context, reserves for

I. U. Rakhmonov et al., *Electric Consumption by Industrial Enterprises*, Power Systems, https://doi.org/10.1007/978-3-031-62676-0_11

electrical power saving (EPS) are identified through the optimization of a complex of indicators for raw materials, materials, and energy.

Consider the general case where a plant of a specific enterprise manufactures products from raw materials and semi-finished products in a certain technological sequence for the manufacturing of the final product Z for this production.

To assess the influence of the material intensity of the final product, i.e., the specific consumption of raw materials and semi-finished products per unit of final product q, we find the value of this indicator before implementing measures to reduce material consumption.

$$q_1 = \frac{V}{Z}, \tag{11.1}$$

and, then:

$$q_2 = \frac{V - \Delta V}{Z}. \tag{11.2}$$

Accordingly, the specific material savings will be:

$$\Delta q = q_1 - q_2, \tag{11.3}$$

and, the change in SEPC per unit of workshop production d based on the standard characteristic $d = f(P)$:

$$d \pm \Delta d, \tag{11.4}$$

where $\pm\ \Delta d$ is the deviation of special consumption. Here, the "+" sign denotes a decreasing standard characteristic in the working zone of productivity (hyperbolic curve), and "–" denotes an increasing dependency (parabolic curve), see Fig. 11.1.

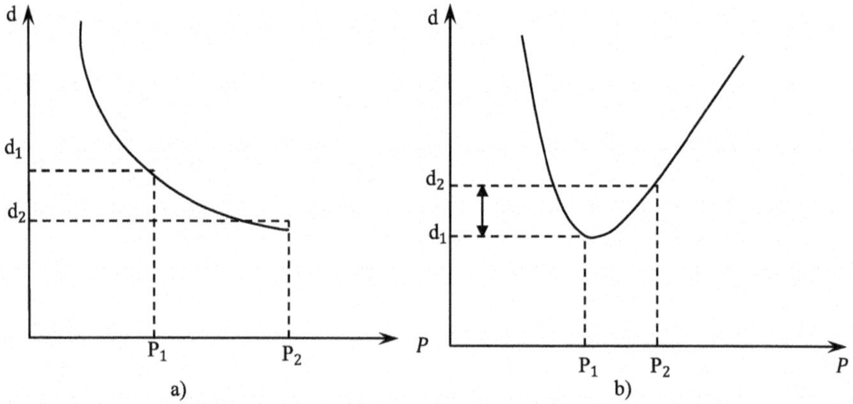

Fig. 11.1 Standard characteristics: **a** parabolic and **b** hyperbolic

If the workshop's proportional share in the SEPC for the final product of the enterprise is similarly:

$$\alpha = dq, \tag{11.5}$$

then the reduction of this share due to the decrease in material intensity will be equal to:

$$\Delta\alpha = dq - (d \pm \Delta d)(q - \Delta q), \tag{11.6}$$

the amount of EPS will be:

$$\Delta W = \Delta\alpha Z, \tag{11.7}$$

and the characteristic standard characteristic of the production object will have the following form:

$$\Delta W = \left[dq - (d \pm \Delta d)(q - \Delta q)\right] \cdot Z. \tag{11.8}$$

If, according to the standard characteristic Δd, the increment has a positive sign $(+\,\Delta d)$, then the amount of savings will be:

$$\Delta W = \left[\Delta q(d + \Delta d) - \Delta dq\right] \cdot Z. \tag{11.9}$$

With a negative sign $(-\,\Delta d)$:

$$\Delta W\left[\Delta q(d - \Delta d) + \Delta dq\right] \cdot Z. \tag{11.10}$$

Thus, the first part of Eqs. (11.9) and (11.10) represents the savings in electrical power due to the reduction in material intensity, increased or decreased depending on the form of the standard characteristic curve; the second part of these equations—partial savings or excess EPC in connection with the change in the volume of consumed raw materials or produced semi-finished products [2].

Equations (11.9) and (11.10) correspond to the most common cases when the electrical power intensity of the finished product decreases, and the volume of its production remains unchanged with the reduction in the production of semi-finished products and the consumption of raw materials. The reduction in EPC can be reflected as:

$$q_2 < q_1; \quad Z = const; \quad P\text{—}var; \quad W\text{—}var. \tag{11.11}$$

These same equations can also be applicable to a number of other industries, such as the textile industry, where the EPC remains constant with changes in the volume of semi-finished products manufactured:

$$q_2 < q_1; \quad Z\text{—}var; \quad P\text{—}var; \quad W = const. \tag{11.12}$$

Completely different values are obtained when the output of finished products can be increased while maintaining the same volume of semi-finished products produced and the same EPC, i.e.:

$$q_2 < q_1; \quad Z{-}var; \quad V = const; \quad W{-}var. \tag{11.13}$$

From formulas (11.9) and (11.10) with $\Delta d = 0$, we obtain:

$$\Delta W = \Delta q d Z. \tag{11.14}$$

Here, the saving of electrical power is directly proportional to the degree of reduction in material intensity.

With a constant volume of finished product output, a reduction in the quantity of semi-finished products manufactured, and a constant EPC, i.e., under the condition:

$$q_2 < q_1; \quad Z = const; \quad V{-}var; \quad W = const, \tag{11.15}$$

we will have:

$$d_1 q_1 = d_2 q_2. \tag{11.16}$$

As the EPC, does not change with the variation in q, or

$$d_1 q = (d_1 \pm \Delta d)(q_1 - \Delta q). \tag{11.17}$$

By substituting this expression into Eq. (11.8), we obtain:

$$\Delta W = 0. \tag{11.18}$$

Thus, if the reduction in material intensity does not decrease EPC and does not change the volume of final production, then this measure will not lead to EPS.

The total EPS for the enterprise as a whole will be:

$$W_э = Z \sum_{i=1}^{n} d_i - (d_i + \Delta d_i)(q_i - \Delta q_i) \tag{11.19}$$

or

$$W_i = \sum_{i=1}^{n} \Delta W_i = Z \sum_{i=1}^{n} \Delta \alpha_i. \tag{11.20}$$

The total EPS at the enterprise due to the reduction in material intensity, considering the energy consumption of producing raw materials and semi-finished products manufactured at other enterprises, can be determined by the method proposed below, which takes into account two most characteristic situations [3]:

(a) a reduction in the volume of raw materials and semi-finished products supplied to the enterprise while maintaining the same quantity of finished products manufactured by the enterprise;
(b) an increase in the output of finished products with the same volume of raw materials and semi-finished products supplied, due to a reduction in their specific consumption.

In the first case, with a reduction in the supply of raw materials and semi-finished products by an amount ΔV while maintaining the same volume of finished product output, the EPS are determined as follows.

Find the SEPC of raw materials or semi-finished products entering production:

$$d_a = \frac{eZ}{V} = e\frac{1}{q}, \tag{11.21}$$

where e—the energy intensity of the raw materials and semi-finished products entering production.

The total EPS from reducing the amount of V by ΔV will be:

$$W_e = e\frac{\Delta V}{q} + \sum_1^n \Delta W \tag{11.22}$$

or

$$W_e = d_a \Delta V + \sum_1^n \Delta W_i. \tag{11.23}$$

In the second case, when the output of the final product is increased by an amount ΔZ while maintaining the same V, the total EPS will be:

$$W_e = e\Delta Z \tag{11.24}$$

When the enterprise receives several types of raw materials m and semi-finished products V_A, $V_B + \cdots + V_m$, the EPS are equal to W_e. If, when reducing supplies (RS) by the amount of $\Delta V_A, \Delta V_B, \ldots, \Delta V_m$ the volume of finished product Z remains unchanged, then:

$$W_e' = \sum_1^m e_i \frac{\Delta V_i}{q_i} + \sum_1^n \Delta W_i', \tag{11.25}$$

or

$$W_e' = \sum_1^m e_i \frac{\Delta V_i}{q_i} + \sum_1^n \Delta W_i'. \tag{11.26}$$

The value of $\Delta W'$ for this case is determined in accordance with the formula:

$$\Delta W' = \Delta\alpha_i' \cdot Z, \tag{11.27}$$

where $\Delta\alpha' = d_1 f - (d_1 \pm \Delta d)\left[\left(q_{A_1} - \Delta q_A\right) + \left(q_{B_1} - \Delta q_B\right) + \left(q_{m_1} - \Delta q_m\right)\right]$, or

$$\Delta\alpha' = d_1 f(d_1 \pm \Delta\alpha)\sum_1^m \left(q_{i_1} - \Delta q_i\right), \tag{11.28}$$

where

$$f = q_A + q_B + \cdots + q_m. \tag{11.29}$$

If the output of finished products increases by an amount ΔZ while maintaining the same RS, then:

$$W_e' = \Delta Z \sum_1^m e_i \tag{11.30}$$

11.2 Selection of Optimal Criteria for Energy Indicators of Energy-Intensive Units and Mass-Use Units

As noted earlier, the main consumer of electrical energy in technologically advanced countries is the industry. Organizing ES and IEE in the industrial sector is not only a current issue but also an extremely complex task. This complexity is due to the great diversity of enterprises in terms of the nomenclature and volumes of products manufactured, as well as the specifics of each technological unit and process [4].

Additionally, differences in the operating regime of each enterprise should be considered—round-the-clock continuous operation, operation with technological breaks, single and double shift operations. This explains the need to solve a number of tasks to reduce EPC, involving a large complex of measures that take into account the features of each electrical power-consuming object.

Based on the results of conducted research and analysis of operational materials on the rational use of electrical power, the following tasks for reducing energy costs can be formulated, covering almost all industrial sectors:

Rationalization of the EPC regime with the solution of the following subtasks:

(a) reducing energy costs in electric power continuous and cyclic production processes;
(b) the same for electrical units of direct and indirect action, high-temperature and low-temperature, as well as continuous and cyclic electrothermal processes;

(c) reducing EPC in unsteady modes of unit operation (start-up and shutdown);
(d) reducing the SEPC when alternating between full power operation, reduced load mode, and idling;
(e) reducing the SEPC during linear and non-linear changes depending on the type of unit and the nature of the load change.

Regulation of electrical loads at all levels of production with the aim of reducing the 'peak' load of the power system:

(a) reducing 'peak' load without additional capital expenditures and without reducing productivity;
(b) the same, but with additional capital investments and reduced productivity;
(c) forming the cyclic nature of technological processes to discretely coincide pauses in operations with the time of peak load in the power system;
(d) creating conditions for stable operation of the electrical load regulation system.

Estimation of the forecasted value of the reduction in "peak" power and its lead time when using technical controls:

(a) **estimation** of the forecasted amount of power reduction (theoretical forecast);
(b) **estimation** of the time ahead of a 30 min excess of declared power during peak hours of the power system.

Reduction of energy consumption through the implementation of technological measures:

(a) optimization of technological parameters in electric power processes;
(b) the same, but in electrothermal processes.

Reduction of electrical power losses in the power supply system, in machines, and apparatus:

(a) reduction electrical power losses in the power supply system;
(b) the same in electrical machines and apparatus, electrothermal units;
(c) optimizing the operation mode of reactive power compensation devices.

Modernization and updating of machinery and equipment to reduce energy consumption:

(a) modernization of the existing fleet of machines and mechanisms;
(b) replacement of machines, mechanisms, processes with less energy-intensive ones.

Implementation of organizational measures to reduce energy consumption:

(a) implementation of an automated system for accounting, control, and management of the consumption of electrical power, compressed air, water, gas, steam, etc.;
(b) organization of the collection and storage of statistical information on the dynamics of energy indicators;

(c) development of standardization methods and organization of control over their compliance.

Thus, the entire complex of research and implementation of ES and IEE measures comes down to solving 8 tasks, which in turn consist of 21 sub-tasks.

Optimization of Electrical Load Modes of Energy-Intensive Units Operating in Various Modes

Ensuring a minimum level of power consumption is one of the most important tasks solved in the development of technology for ES and IEE industrial facilities. It should be kept in mind that the nature of power consumption strictly adheres to the technological regulations of the unit operating in a mode of both constant and variable load. The complexity of solving the task also lies in the fact that in operating conditions, there is a significant deviation of electrical loads due to random factors [5].

The use of a set of "typical graphs" of electrical loads and their probabilistic characteristics does not provide the desired accuracy in calculating electrical loads and their optimization under the operating conditions of industrial energy equipment, as the difference between calculated and actual loads reaches 200%.

Research conducted using a systems approach allows for the development of methods for optimizing electrical loads by determining clear dependencies between the operation of the electric drive and their energy indicators.

The solution to the set task for the most characteristic operating conditions of the units is considered below.

Continuous Operation Mode of the Unit

This mode can be of two types:

(a) continuous operation mode with constant hourly productivity;
(b) continuous operation mode with varying productivity.

When units operate in modes a and b, the consumed power can be expressed by the following formula:

$$P = P_{n.l} + P_{aux} + P_u, \tag{11.31}$$

where $P_{n.l}$ is the no-load power,

P_{aux} is the power of auxiliary mechanisms,

P_u is the useful power of the unit.

In cases where, under the influence of random factors, there is a fluctuation in the value of P, the average value P_{avg} is taken as the calculated load; powers $P_{n.l}$ and P_{aux} can be considered constant. An exception can be made for units whose $P_{n.l}$ may undergo certain changes during the calculated period. An example of this can be cement tube ball mills, the grinding bodies of which lose weight over time, significantly reducing the consumed no-load power.

Continuous Operation Mode of the Unit with Constant Hourly Productivity

In this mode, the consumed power P is practically constant provided that: $P_{n.l} = const$; $P_{aux} = const$; $P_u = const$; $A = const$.

Under the condition $P_{n.l} = var$, and $P_{aux} = const$, the resultant power for the calculation period changes within certain limits.

Continuous Operation Mode of the Unit with Varying Productivity

In operating conditions, various disruptions in the rhythm of the supply of products to be processed may occur, which leads to changes in productivity and, consequently, consumed power. In this case, the nature of the change in consumed power can be expressed by formula (11.31).

Operation of the Unit in a Variable Mode

Units can operate in several modes:

(1) Alternating maximum load with complete shutdown of the unit.
(2) Alternating maximum load with the unit's idle operation.
(3) Alternating maximum and reduced loads.

In all modes, the average power is taken into account, which depends on the productivity $P_{avg} = f(A_{avg})$.

A special place when calculating the power consumption in the unit when operating in the indicated modes is occupied by the energy consumption of the units during their start-up, which in some cases have a significant share in the total EPC. The magnitude of starting expenses (startup and shutdown mode) significantly influences the energy indicators of the unit, and each has its peculiarities. The magnitude of these expenses is mostly determined experimentally and should be considered in the process of optimizing EPC modes.

11.3 Main Directions for Reducing Electrical Power Losses

At industrial enterprises, rational use of electrical power can be ensured, among other things, by optimal distribution of electrical power between workshops and other structures of the enterprise, taking into account the nature of the technological process.

For making adequate decisions to improve energy efficiency, it is necessary to have the ability to forecast energy consumption at different times for various production conditions. It is also essential to address the question: in which departments and stages of the technological process can the maximum positive effect be achieved?

The development and refinement of methods for determining SEPC indicators and increasing the accuracy of calculations, as well as the practical use of the results discussed in this chapter, allow for identifying points and reserves of energy savings and forming plans for measures to increase energy efficiency.

Based on the conducted research on the rational use of electrical power, measures to reduce energy consumption at industrial enterprises have been proposed, implementable in four directions.

These measures are based on the results of developing new methods and technical solutions for reducing EPC in the production process, as well as on mathematical models and algorithms for managing EPC created for its minimization [6, 7]. Simultaneously, technological and operational factors affecting the reduction of the energy intensity of manufactured products were investigated.

From the research results for specific economic sectors, a unified information fund can also be created, reflecting the technical, technological, organizational, and economic aspects of transformation, transmission, distribution, and EPC. In this context, the comparison of energy efficiency of consumption is carried out only under identical or similar production conditions.

Enterprises in the metallurgical industry are characterized by similar productions and practically identical equipment. This allows not only comparing their EPC efficiency but also identifying factors that led to deviations in energy efficiency indicators at some of them, as well as the magnitude of these deviations.

The directions and measures implemented within each of them, Fig. 11.2 are acceptable and effective for all industrial enterprises.

Analysis of the measures carried out in the direction of "Reducing electrical power losses", Fig. 11.3 shows that their implementation allows for a reduction in the enterprise's energy consumption by an average of 3–5%.

For instance, ensuring the nominal voltage value in ferrous metallurgy during steel smelting significantly reduces power losses. Moreover, 80% of the equipment at industrial enterprises is the main technological equipment of the production process. Quality current and intermediate maintenance of this equipment not only reduce EPC but also extend its service life.

Implementing the measures shown in Fig. 11.4 at an industrial enterprise can reduce EPC by an average of 5%. Here, the main factor is the quality of the primary

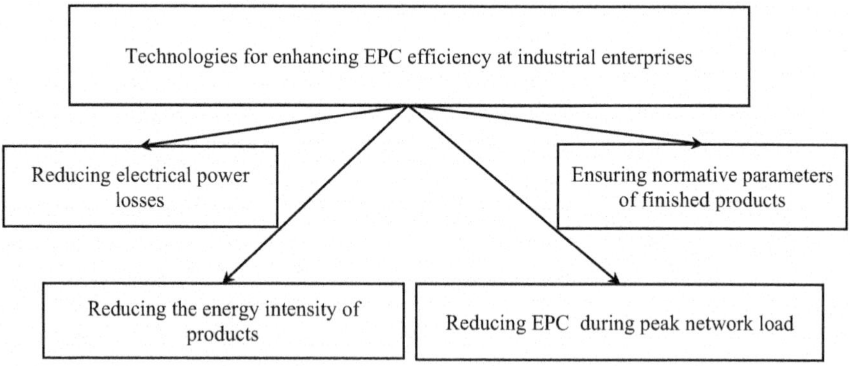

Fig. 11.2 Directions of technology for rational EPC at industrial enterprises

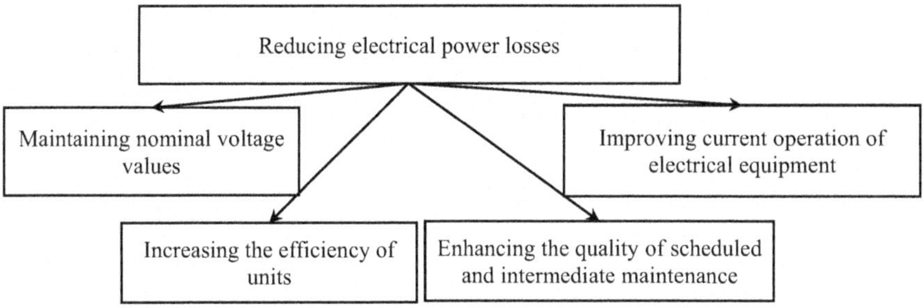

Fig. 11.3 Measures carried out to reduce electrical power losses

raw materials. Research conducted in metallurgical, cotton cleaning, and textile enterprises shows that establishing normative indicators of the quality of primary raw materials contributes to ensuring normative EPC.

In metallurgy, there is a discrepancy between EPC and the composition of the primary charge, and in cotton cleaning enterprises (or sectors), between the composition and quality of raw cotton. In other words, the deterioration in the quality of primary products (raw materials) leads to increased EPC. Quality control of raw materials is a critical stage in the technological process. Reducing EPC for auxiliary procedures carried out in the technological process leads to a decrease in the energy intensity of products [8].

As a result of implementing measures to ensure normative parameters of finished products, an average EPS of 2–4% can be achieved (Fig. 11.5). Among these measures, maintaining the normative temperature plays a crucial role. For example, the inability of cotton cleaning plants to maintain the normative temperature during cotton storage affects its grading, which, in turn, leads to increased energy consumption during its processing.

It is known that industrial enterprises pay for electrical power based on a differentiated tariff. Implementing measures to reduce EPC during peak energy consumption

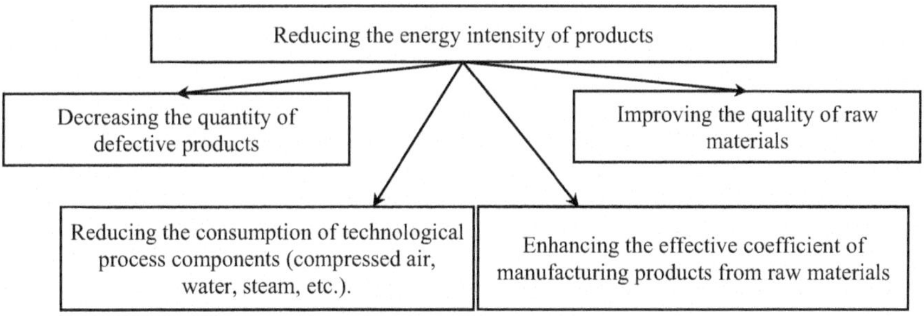

Fig. 11.4 Measures to reduce the energy intensity of products

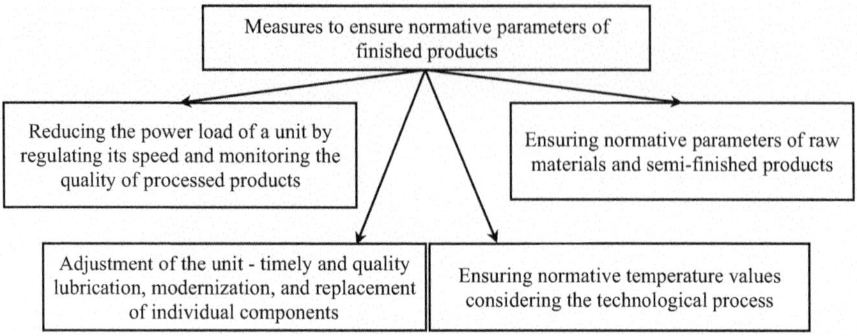

Fig. 11.5 Measures to ensure normative parameters of finished products

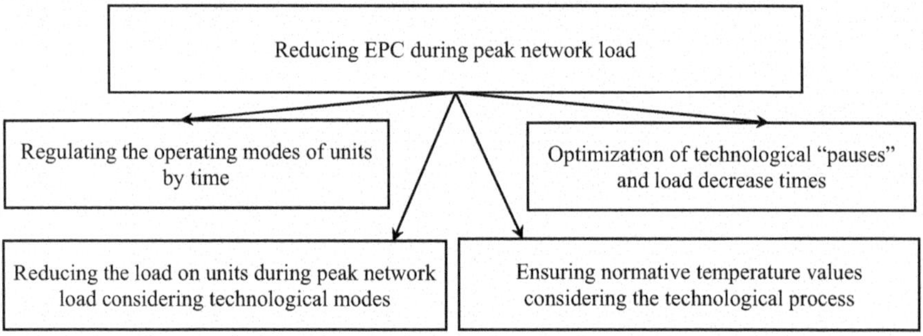

Fig. 11.6 Measures to reduce EPC during peak network load

(in the morning and evening hours) (Fig. 11.6) leads to a reduction in the average daily electrical power bill. This, in turn, reduces the cost of production and enhances its competitiveness.

References

1. IEA (2021) Driving energy efficiency in heavy industries—analysis. Retrieved from https://www.iea.org/reports/driving-energy-efficiency-in-heavy-industries
2. Roelofsen O, Somers K, Speelman E, Witteveen M (2020) Plugging in: what electrification can do for industry. McKinsey Company. Retrieved from https://www.mckinsey.com/business-functions/sustainability/our-insights/how-electrification-can-help-industrial-companies-cut-costs
3. Tomonobu S, Vikas K (2023) Analysis of the efficiency of energy management at the metallurgical enterprise. Econ Dev Environ Ecosyst 9(8):14–26. https://doi.org/10.1007/978-3-031-26596-9_8
4. Nataliia D, Tatiana B, Igor Z (2021) Energy efficiency of entrepreneurship as a marketing procedure and a way to increase efficiency of enterprises, 2021. Mark Infrastruct. https://doi.org/10.32843/infrastruct60-13

5. Rakhmonov IU, Ushakov VYa, Niyozov NN, Kurbonov NN, Mamutov M (2021) Energy saving in industry. E3S Web Conf 289:07014. In: Energy systems research 2021. https://doi.org/10.1051/e3sconf/202128907014
6. Rakhmonov IU, Ushakov VYa, Najimova AM, Jalilova DA, Omonov FB (2021) Minimization of energy production management consumptions. E3S Web Conf 289:07013. In: Energy systems research 2021. https://doi.org/10.1051/e3sconf/202128907013
7. Rakhmonov IU, Kurbonov NN (2020) Analysis of automated software for monitoring energy consumption and efficiency of industrial enterprises. E3S Web Conf 216:165717. In: Rudenko international conference on methodological problems in reliability study of large energy systems. https://doi.org/10.1051/e3sconf/202021601178
8. Hoshimov FA, Rakhmonov IU, Niyozov NN (2020) Technology to reduce energy costs in the electric steel melting shop. E3S Web Conf 209:07017. In: Sustainable development and smart management, ENERGY-21. https://doi.org/10.1051/e3sconf/202020907017

Conclusion

A complex multi-parametric task of exact forecasting the electricity consumption of industrial enterprises for realizing the potential of energy saving and enhancing energy efficiency was successfully solved. It is important to consider several crucial aspects: the peculiarities of the technological process, the operating modes of electrical equipment, and the rationing and forecasting of electricity consumption using modern high-precision methods.

In the practical methods of EPC management, the main criterion is to ensure the normative values of technical and economic indicators, i.e., the management method is aimed only at achieving the minimum/maximum of a given indicator. For high energy efficiency, it is necessary to manage electricity consumption by combining these indicators.

The monograph confirms the feasibility of managing the EPC of industrial enterprises using modeling, standardization, and forecasting. This required, among other things, the creation of a comprehensive classification of EPC objects, covering both main and auxiliary divisions of enterprises.

In modern market conditions, the rational use of energy resources and minimizing energy consumption in the technological process significantly affect the level of net cost and the price of the final product, and ultimately—the competitiveness ("viability") of the enterprise. For energy supplying companies, the positive effects include increased stability of operating modes, reduced net cost of electricity, and improved technical and economic indicators.

The monograph emphasizes the importance of strictly adhering to the norms of specific electricity consumption of energy resources, their timely correction for the future, considering possible changes in production capacities and the improvement of technological processes. It is necessary to estimate the potential for energy savings at an accelerated pace.

A scientifically substantiated set of energy-saving measures for industrial enterprises of various sectors of the economy is proposed, considering the peculiarities of their technological processes. It is shown that rational energy use measures should

I. U. Rakhmonov et al., *Electric Consumption by Industrial Enterprises*, Power Systems, https://doi.org/10.1007/978-3-031-62676-0

be carried out in four directions: reducing electricity losses, reducing the specific energy intensity of products, reducing EPC during peak loads in the electrical grid, and ensuring normative parameters of the finished products.

The authors hope that the methods, algorithms, and models for standardizing and forecasting EPC, tested on a number of industrial enterprises, will allow managers and specialists to identify reserves and significantly improve the competitiveness of the products manufactured by their enterprise.

The manufacturer's authorised representative in the EU is Springer Nature Customer Service Centre GmbH, Europaplatz 3, 69115 Heidelberg, Germany. If you have any concerns regarding our products, please contact ProductSafety@springernature.com

Printed and bound by CPI Group (UK) Ltd, Croydon, CR0 4YY
01/10/2026
02225987-0010